T0254049

This volume provides a detailed country-by-country account of the increase in forest resources in Europe over the past forty years. This expansion, in standing volume and to a lesser extent in area, is a continuation of a trend that began during the nineteenth century. After presenting the historical developments, the author discusses the implications, should this trend be allowed to continue, for the future health and vitality of the forests, for forest policy management and silviculture, and for the economic viability and environmental sustainability of the resource. An increase in thinnings and regeneration cuttings is advocated, replacing current unstable tree species by true climatic climax species, as is a shortening of the currently over-long rotation ages. The author concludes that preserving the sustainability and biodiversity of Europe's forest ecosystems can be achieved by maintaining the genetic diversity, density, age and health stability of forests, protecting biotopes of endangered species and establishing cultural biotopes and strictly protected natural reserves.

FOREST RESOURCES IN EUROPE 1950–1990

FOREST RESOURCES IN EUROPE 1950–1990

KULLERVO KUUSELA

CAMBRIDGE
UNIVERSITY PRESS

CAMBRIDGE UNIVERSITY PRESS
Cambridge, New York, Melbourne, Madrid, Cape Town, Singapore, São Paulo

Cambridge University Press
The Edinburgh Building, Cambridge CB2 8RU, UK

Published in the United States of America by Cambridge University Press, New York

www.cambridge.org
Information on this title: www.cambridge.org/9780521480765

First published 1994
Reprinted 1996
This digitally printed version 2008

A catalogue record for this publication is available from the British Library

ISBN 978-0-521-48076-5 hardback
ISBN 978-0-521-05223-8 paperback

The views expressed in this study are the author's and do not
necessarily correspond with those of the European Forest Institute.

CONTENTS

FOREWORD

The Timber Section of the ECE/FAO Agricultural and Timber Division has prepared five decennial assessments of the European forest resources. The last one, dated 1990, includes the industrial countries of the Temperate zone. The assessments are based on the information provided by the countries and they form an indispensable information basis for developing forestry - in its all forms - and forest industries in Europe.

The European Forest Institute started its activities in the beginning of 1993. Already in the planning phase of EFI the discussions with the Timber Section of the ECE/FAO Agricultural and Timber Division led to the conclusion that an independent institute could initiate a study based on all ECE/FAO Forest Resource Assessments. The cooperation between EFI and the Timber Section in developing a computerized database using the published assessments made it possible to produce the basic material for this kind of study.

European Forest Institute entrusted the work to Professor Kullervo Kuusela, leader of the Finnish National Forest Inventory for almost 30 years. He has also prepared the Finnish reports for the ECE/FAO Forest Resource Assessments 1970 and 1980, and contributed to the forest resource forecasts as a part of the 1980 Assessment. His earlier international work is focused on the study *European Forest Resources and the Trade of Industrial Wood in 1950-2000*, which was published in Finnish in 1985.

This study is the first issue in EFI's research report series. It consists of four main parts: introduction, wood resources and harvests by country groups, country statistics and conclusions on the ecological and economic basis of forestry in Europe. The scope of the study is a large one, and many of the observations and conclusions presented by Professor Kuusela are of high interest regarding the development of European forest resources during the past forty years. I am convinced that this kind of critical assessment of the medium-term development trends of forest resources is needed in economically uniting Europe.

It is my pleasure to thank Mr Tim Peck and Mr Klaus Janz for reading the manuscript and for the valuable comments they have made, and the Timber Section of the UN-ECE/FAO Agricultural and Timber Division for providing the data and for good cooperation. I wish to express special thanks to the forest inventory experts in the countries within the scope of this study for their comments on the observations made on their countries in Chapter 3. I also wish to acknowledge Mr Ashley Selby and Ms. Leena Roihuvuo for editing the text, and Mr Juho Pitkänen and Ari Turkia, who prepared the tables and figures for the report.

Birger Solberg, Director
European Forest Institute
Joensuu, May 1994

ABBREVIATIONS

Term	Symbol
Forest lands (forest and other wooded land)	FLs
Forest Land on which grows:	
Forest (closed forest)	
Exploitable forest	EF
Unexploitable forest	UEF
Other wooded land	OWL
Coniferous percentage	C-%
Forecast of resource variable for 1990 made in 1980. Forecast is the average of high and low values presented in the 1980 Assessment	FC90
Growing stock	GS
Net annual increment	NAI
Natural losses	NL
Gross annual increment	GAI
Increment percentage of growing stock	I-%
Fellings percentage of growing stock	F-%
Drain	D
Fellings	F
Annual fellings	AF
Logging residues	LR
Removals over bark	Rob
Removals under bark	Rub
Bark	B
Recorded (inventoried) growing stock	RGS
Calculated growing stock at the end of a given period, estimated on the basis of growing stock at the beginning of the period; and increment and drain during the period	CGS
Gross annual increment during a 10-year period	10GAI
Net annual increment during a 10-year period	10NAI
Fellings during a 10-year period	10AF
Drain during a 10-year period	10AD
Ratio CGS/RGS	C/R

FIGURES

TABLES

SUMMARY

The accuracy and consistency of the statistics of forest resources, fellings and natural losses reported by countries for use in the UN-ECE/FAO Forest Resource Assessments are insufficient with regard to their importance as a basis for estimating forest policy implications and for developing policies and management regimes for multi-benefit forestry in a Europe that is becoming economically and politically more integrated.

A way to improve the situation would be a system of repeated sampling measurements and observations by countries covering all relevant forest resource variables, including those concerning the health and multi-benefit management of forests. Sampling units should be tied to time and space coordinates, which permit the calculation of results for ecological and economic regions regardless of national boundaries.

Wood harvesting and utilization statistics should be of such a quality that reliable forest balance estimates are possible. The weakest point in the current utilization statistics concerns the amount of wood used for fuel and household purposes. Moreover, the definitions employed should be consistent so that country results would be comparable.

Analyses concerning the genetic, density, age and health situation and development trends of forests should be made with reference to the serial successions and climatic climaxes of forest plant communities under the effects of changing environmental factors.

In order to obtain a reliable assessment of the economic profitability of forestry, which has been decreasing and is negative in large areas, income and cost analyses should be made and applied to the rationalization and modernization of working methods, equipment and machinery, and to develop a pricing and valuation system for all material and non-material forest benefits, as well as a system for financing sustainable multi-benefit forestry.

Europe's recorded growing stock increased by 43% during the period from 1950 to 1990 and is now increasing faster than ever. The net annual increment of the growing stock is about 175 million m³, over bark, greater than the annual fellings. The difference is about 145 million m³, expressed in under bark units of removals.

If this trend continues, stand density and age and growing stock volume per hectare will increase and reach the stage at which biological stability begins to deteriorate with increasing insect, fungal and wind damage and other natural losses. The risk is greatest in those areas of planted coniferous stands on the sites where oak, beech and other broadleaved trees are the natural climatic climax species, and where thinnings have been postponed. The bulk of those coniferous stands are older than 120 years, and the average growing stock volume is c. 300 to 400 m³ per hectare.

The biological degradation, if permitted to continue uninterrupted, will cause considerable economic losses and environmental impoverishment.

These trends can only be stopped and the stability of forests maintained by increasing thinnings and regeneration cuttings, and by replacing the current unstable tree species by true climatic climax species, as well as by shortening the current over-long rotation ages. The reduction of pollution emissions should also be continued.

If these objectives are accepted, the sustainable fellings could equal the current net increment. In other words, the recorded current annual removals in Europe, 369 million m³ under bark, could be increased by about 145 million m³. The increase of removals could make Europe a net exporter of forest products, unless the consumption of wood were to be much increased. Reaching this target would require revolutionary changes in forest management regimes in many countries.

On the consumption side, the capacity of forest industry should be increased. This concerns especially chemical pulping and the production of wood-based panels, and the use of wood and its residues for energy and in buildings and furniture.

The area of other wooded land in Europe is 46 million ha, of which 36 million ha are located in the Mediterranean region. The corresponding areas of unexploitable forest are 16 and 8 million ha respectively. In the Mediterranean, the present scrub and low-density tree and woody vegetation has been used as the criterion for separating other wooded land from forest and exploitable forest. A great part of other wooded land and unexploitable forest land could be improved and afforested to grow closed forest. Europe also has surplus agricultural land amounting to 15 to 20 million ha.

The area of the European closed forest could be increased by at least 40 million ha if an increase in forest and tree biomass were to be agreed upon as a policy objective.

From the point of view of land use policy aiming at the full utilization of the potential of forest lands, the worst obstacles are the low standards of living and a high proportion of agricultural population in many areas, and the low rates of industrialization and disintegrated economic and social structures in the former centrally planned economies, as well as antagonistic attitudes towards forestry proper.

Forestry can help support sustainable development via afforestation and forest improvements which increase tree and biomass, decrease the net emission of carbon dioxide into the atmosphere. Wood grown sustainably and used in all branches of consumption and energy production provides an additional way to replace fossil materials.

Preserving the functioning and biodiversity of forest ecosystems as another objective of sustainable development can be reached by maintaining the genetic, density, age and health stability of the forests, by protecting biotopes of endangered species and by establishing cultural biotopes and strict natural reserves. Man's ethical responsibility must be enlarged to encompass the whole biosphere and the survival of all plant and animal species.

1 INTRODUCTION

1.1 General background

European forests and forestry are at a stage in which both the current situation and development trends need to be analysed and the policy implications formulated. This can be illustrated by the following observations.

Wood resources have increased constantly since the 1950s. They are greater now than at any time during the last 200-300 years. This has been achieved mainly by the management regimes developed and implemented in the 19th and 20th centuries. Moreover, as a result of the considerable developments in energy, production and traffic technologies, the use of wood for fuel and in buildings and other construction as well as utility commodities has diminished. Because of the decrease of the agricultural population, the use of wood for household purposes has also been reduced. At the same time, the increased efficiency of agricultural production, including pasturing, has released land for alternative uses, including forestry.

Side by side with the relatively decreasing importance of the commodity function of forests, protective, environmental, social and cultural functions have gained importance. Rising standards of living and purchasing power have given rise to wood substitution. The forest has become more of an environmental and social asset than a commodity asset in many countries.

The vitality of forests has become a dominating topic in discussions concerning forests and forestry practices. In addition to wild fires, increasing emissions from industry, traffic and consumption residues and their depositions in forests and other parts of the biosphere have killed trees and stands on some sites. They are assumed to contribute to a large-scale loss of vitality in forests.

Globally, the human population explosion, the reckless exploitation of forests and other natural resources, the accumulation of anthropogenic emissions in the biosphere and the increasing amount of carbon dioxide and other greenhouse gases in the atmosphere are increasingly believed to damage the functioning of ecosystems, to decrease biodiversity, and, consequently, to weaken the basis of man's life on the earth.

These trends and threats have led to efforts to seek international agreements and co-operation towards achieving sustainable development, monitoring and managing natural resources in balance with the necessary satisfaction of physical and cultural needs. Properly functioning ecosystems, rich biodiversity and a healthy living and working environment are recognised to be essential goals.

Forests have a central role in the strategy and implementation of sustainable development. They are the largest and the most diversified terrestrial ecosystem. Although the current area of European forests is hardly a third of its original extension, forests have been disturbed relatively little by human activities compared with fields, pastures and gardens.

Growing and using wood is a part of the natural flow of solar energy and the circulation of atmospheric carbon. Tree biomass is created through photosynthesis, which fixes carbon dioxide and releases oxygen. Under natural conditions, wood eventually decomposes, consuming oxygen and releasing carbon dioxide. The latter returns to the biosphere and is once again fixed in photosynthesis. When man uses wood for fuel, and commodities decay or are burnt, carbon is released into atmosphere and back into its natural circulation.

The more man substitutes fossil materials by sustainably produced wood, the greater the decrease in the net emissions of carbon dioxide. Additionally, the consequent increase of the forest biomass absorbs more carbon dioxide, further reducing the net emission.

A successful forest production regime which satisfies human needs and supports sustainable development can only be based on reliable and thoroughly analysed resource information. Forestry activities must be adjusted to the ecological, economic and historical conditions of different countries and country groups.

1.2 *Information base*

The basic information for this study consists of the Forest Resource Assessments of Europe made by the UN-ECE Timber Committee and FAO European Forestry Commission and their joint secretariat, the Timber Section of the Joint FAO/ECE Agricultural and Timber Division in Geneva, for the years 1950, 1960, 1970, 1980 and 1990. During this 40-year period the contents and definitions of resource characteristics have been developed towards a more consistent system of concepts and terms. Especially in the first Assessments, but also in the later ones, the concepts used by the reporting countries have differed from each other to such an extent that the consistency of the time series describing the development of forest resources and wood production is not uniform. This concerns, for instance, the concepts *forests in use, exploitable* and *unexploitable forests,* as well as the concepts *growing stock, increment, removals, fellings* and *drain,* which can be presented in cubic metres under or over bark.

The development series from 1950 to 1980 are based on the SITRA (Finnish National Fund for Research and Development) study *European Forest Resources and the Trade of Industrial Wood in 1950-2000,* published in 1985. The study describes the development of forest resources by estimates in which the comparability problem is taken into account as far as possible.

The framework of the ecological and economic bases has been formulated step by step by the author's contributory studies to a number of reports, e.g. the OECD report *The State of Environment,* 1985; the reports concerning European forest resources and timber trends published by the UN-ECE/FAO in 1985 and 1986; an FAO report concerning the development of European forest resources in 1986; and a paper Forest Decline, which is part of the report *Towards Ecological Sustainability in Europe,* undertaken at the International Institute of Applied Systems Analyses (IIASA) in late 1980s. The framework of the ecological basis applied to the boreal zone is presented in the SITRA study *The Dynamics of Boreal Coniferous Forests,* 1990.

The list of source and background literature includes publications and papers considered to be relevant as an information basis of this study. It should be noted that the author has made estimates in order to fill in the gaps in the information base of the UN-ECE/FAO Forest Resource Assessments, and in some cases modified the data in order to get the figures more comparable. Therefore the figures presented in this study are not always identical with those presented in the UN-ECE/FAO statistics.

1.3   *Countries and their groups*

In the UN-ECE/FAO studies of European Timber Trends and Prospects, countries have been grouped on economic and political lines. From the ecological and historical points of view, the forests in these groups are so heterogeneous that the country group averages of characteristics are not very informative. Indeed, they can often be misleading. For example, the mean growing stock volume of EC countries tells little about the mean volumes of Italy, the former Federal Republic of Germany, or

Ireland. Similarly, the tree species composition of these countries is quite different compared with that of Denmark, for example.

In order to improve the consistency of the presentation of forest characteristics within country groups and to clarify regional differences, the present study employs forest vegetation zones as the principal criterion for grouping the countries. Although country boundaries do not follow the boundaries of ecological zones, they considerably enhance the consistence of country group characteristics.

Country groups are presented in Table 1.1 and Fig. 1.1 The ecological basis for grouping is described in Chapter 4. In some groups, countries are at different levels of economic development. The history of forests and management regimes may also differ. The importance of these differences from the point of view of policy implications is analysed in Chapter 4. Iceland and Malta are excluded in the statistics but included in the country summaries. The former Soviet Union is not included in this study, mainly because the data relate to a period prior to the break-up of that country.

Fig. 1.1   *Country groups by ecological zones. Compare Table 1.1*

Table 1.1   *Countries by ecological zones*

I	Coniferous	Ia	Boreal	Ib	Sub-Alpine
I-II	Ecotonal mixed	I-IIa	Latitudinal	I-IIb	Elevational
II	Broadleaved deciduous				
III	Mixed evergreen				

x = principal zone   o = additional zone

	I a	I b	I - II a	I - II b	II	III
Northern						
Finland	x					
Norway	x		o			
Sweden	x		o			
Central						
Denmark			o		x	
Germany West			o	o	x	
Germany East			o	o	x	
Poland			o	o	x	
Czechoslovakia			o	o	x	
Atlantic						
Ireland					x	
United Kingdom				o	x	
Sub-Atlantic						
Netherlands					x	
Belgium					x	
Luxembourg					x	
France		o		o	x	o
Alpic						
Austria		o		o	x	
Switzerland		o		o	x	
Pannonic						
Hungary					x	
Romania				o	x	
Mediterranean West						
Portugal					o	x
Spain		o		o	o	x
Mediterranean Middle						
Italy		o			x	o
Yugoslavia		o			x	o
Albania					x	o
Mediterranean East						
Bulgaria				o	x	o
Greece		o		o		x
Turkey		o		o		x
Cyprus						x
Israel						x

2 WOOD RESOURCES AND HARVESTS BY COUNTRY GROUPS

2.1 Forest lands and their development trends

Forest lands, FLs, are divided into classes as follows.

     Forest (closed forest)
          Exploitable   EF
          Unexploitable   UEF
     Other wooded land   OWL

In principle, forest land represents sites capable of growing closed stands. In Northern Europe, the minimum yield of fully closed stands during rotation is defined to be 1 m³ per hectare and per year. Other wooded land can only grow scattered and stunted trees and bushes. Wetlands, stony mineral sites, and sites above the latitudinal and elevational boundary of closed forests are examples of other wooded land proper. Poor stocking, especially in Mediterranean Europe, is used as an additional criterion in defining the boundary between forest land and other wooded land.

Exploitable forest is primarily managed to serve the wood-producing function. Multi-benefit management, however, is common in large areas of exploitable forest. In unexploitable forests wood production proper is not economically feasible or it is restricted or prohibited because of functions other than production.

The total area of forest lands in Europe, 195 million ha, comprises 77% of forest and 23% other wooded land (Table 2.1). Other wooded land is concentrated in Northern Europe because of climatic and latitudinal effects and also elevational tundra, and in Mediterranean Europe (Table 2.2).

A significant part of the total area of other wooded land in Mediterranean Europe, 36 million ha, is a huge land reserve, should the increasing of forest and its biomass come to be considered a desirable objective.

Exploitable forests amount to 133 million ha, 89%, and unexploitable forests 16 million ha, 11%, of the European forest area. The proportion of unexploitable forests has increased during the recent decades because of the high logging and silvicultural costs in mountainous conditions and in regions of poor climate, and also because of the need to increase the number of protected forests, national parks and nature reserves.

Table 2.1   The forest lands of Europe

	Million ha	%	
Forest	149	77	100
Exploitable	133		89
Unexploitable	16		11
Other wooded land	46	23	
Forest lands	195	100	

Table 2.2   Proportional shares of forest lands by country groups (per cent of European total)

Country group	Forest lands	Forest	OWL	EF	UEF
Northern	31	36	17	36	31
Central	12	16	1	18	5
Atlantic	1	2	0	2	0
Sub-Atlantic	8	9	2	10	4
Alpic	3	3	0	3	4
Pannonic	4	5	0	5	7
Mediterranean W.	15	8	39	7	14
Mediterranean M.	10	11	7	10	19
Mediterranean E.	16	10	34	9	16
Europe	100	100	100	100	100

Poor stocking is an additional criterion for defining the boundary between exploitable and unexploitable forests in the Mediterranean area. The unexploitable forest area amounts to 8 million ha, and this could be also a land reserve for increasing forest resources.

The trends of the classes of forest lands during recent decades (Table 2.3 and Fig. 2.1) are obscured by the changing definitions in compiling area statistics. In Northern Europe, other wooded land increased from 6.6 million ha in 1960/70 to 7.9 million ha in 1980 in spite of wetland drainage. Greater changes have occurred in Mediterranean Europe because of the heterogeneous forest conditions and the lack of sampling inventories. Thus, the area of forest decreased from 12.5 million ha in 1960/70 to 9.5 million ha in 1980, whereas the area of other wooded land increased from 15.5 million ha in 1980 to 28.8 million ha in 1990 in Mediterranean West. In the latter case this was mainly due to a reclassification of land categories in Spain.

Table 2.3   **Recorded forest, other wooded land and their totals in the 1960s and early 1970s, 1980 and 1990, by country groups, million ha**

Country group	Forest			Other wooded land			Forest lands		
	1960/70	1980	1990	1960/70	1980	1990	1960/70	1980	1990
Northern	51.2	51.9	53.2	6.6	7.9	7.7	57.8	59.8	60.9
Central	23.3	23.2	24.1	0.3	0.7	0.2	23.6	23.9	24.3
Atlantic	1.9	2.3	2.6	0.3	0.2	0.2	2.2	2.5	2.8
Sub-Atlantic	14.6	14.9	14.2	0.0	1.4	1.0	14.6	16.3	15.2
Alpic	4.6	4.7	5.0	0.1	0.2	0.1	4.7	4.9	5.1
Pannonic	7.6	7.5	7.9	0.5	0.4	0.1	8.1	7.9	8.0
Mediterranean W.	12.5	9.5	11.2	4.6	6.0	17.6	17.1	15.5	28.8
Mediterranean M.	14.8	16.4	16.1	3.4	3.4	3.3	18.2	19.8	19.4
Mediterranean E.	14.5	15.1	15.0	15.3	15.1	15.3	29.8	30.2	30.3
Europe	145.0	145.5	149.3	31.1	35.3	45.5	176.1	180.8	194.8

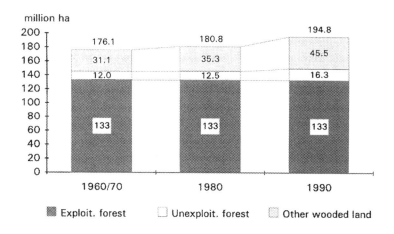

Fig. 2.1   **Estimates of exploitable and unexploitable forest and other wooded land of Europe**

The recorded area of European forests increased from 145.0 million ha in 1960/70 to 149.3 million ha in 1990. If the definitional changes of land classes are taken into account, the actual increase may be even greater. It is consequent upon the vast afforestation of treeless and poorly stocked lands, fields and pastures and by the drainage of wetlands, which have more than compensated for the transfer of forest land into uses other than timber production.

As a result, the net increase of European forests has been at least 5 million ha since 1960/70.

The proportion of forest in the total land area (Table 2.4), is 27% in Europe. It is greatest, 52%, in Northern Europe and smallest, 8%, in Atlantic Europe. The smallest forest percentages are found in the most densely populated countries.

Table 2.4   *Forest as a percentage of the land area, the percentage of coniferous forest, and forest per capita, by country groups*

Country group	Forest-%	C-%	ha/cap
Northern	52	89	2.99
Central	29	70	0.17
Atlantic	8	75	0.04
Sub-Atlantic	23	37	0.17
Alpic	40	74	0.35
Pannonic	25	28	0.23
Mediterranean W.	19	47	0.23
Mediterranean M.	28	24	0.13
Mediterranean E.	14	50	0.18
Europe	27	63	0.26

Before the influence of man, the European forest coverage may have been approximately 80% of the land surface. The decrease has been caused by the clearing of forests for fields, pastures, gardens, towns, industries and infrastructure. The forest coverage is still relatively large in some densely populated countries, e.g. in Alpic and Mediterranean Europe. Forests have survived in mountainous areas outside the centres of population, while the high forest proportion in Alpic Europe is an example of results gained by an efficient long-term land-use policy based on compensating decreases in forest by afforestation.

The proportion of coniferous forests - the proportional area of forest predominated by coniferous trees compared with the total area of forest (Table 2.4) - is 63% in Europe. It is greatest, 89%, in Northern Europe and relatively large in Central, Atlantic and Alpic Europe, respectively 70%, 75% and 74%. Northern Europe is mostly located in the boreal coniferous zone and the tree-species composition there is close to the natural climax. In Central and Atlantic Europe, on the other hand, the large proportion of coniferous species is a result of planting coniferous trees in areas where broadleaved trees, such as oak and beech, are the natural climax trees.

In Alpic Europe, a part of the coniferous stands occur in elevational coniferous and mixed zones. Most of the coniferous stands on lower elevations are planted on broadleaved sites.

In Mediterranean West, planted coniferous stands are quite frequent. The relatively high proportion of coniferous species in Mediterranean East, 50%, is explained by the occurrence of elevational coniferous and mixed forest zones.

The area of forest per capita in Europe is 0.26 ha (Table 2.4). The greater the population density, the smaller the proportion of forest: only 0.04 ha per capita in Atlantic Europe. The largest proportion by far is found in Northern Europe - 2.99 ha per capita. A relatively large area, 0.35 ha per capita, in Alpic Europe demonstrates again the results of an efficient land-use policy.

2.2 Exploitable forest resources 1950-1990

2.2.1 Area

The development of recorded exploitable forest by country groups 1950-1990, its area in 1990 and the area forecast in 1980 for 1990 are presented in Table 2.5 and Fig. 2.2. Forecast (FC90) values are averages of the higher and lower estimates presented in the 1980 Assessment.

Table 2.5   *Development of the area of exploitable forest by country groups, 1950-1990, and the forecast for 1990, made in 1980 (FC90)*

Country group	1950	1960	1970	1980	1990	FC90	Area 1990
	Index: 1980 = 100						1000 ha
Northern	98	92	101	100	100	101	48197
Central	93	96	98	100	103	102	23269
Atlantic	70	78	76	100	110	115	2601
Sub-Atlantic	85	83	98	100	94	101	13493
Alpic	101	97	102	100	108	100	4423
Pannonic	102	85	100	100	92	101	6737
Mediterranean W.	93	100	97	100	97	113	8852
Mediterranean M.	105	104	100	100	96	101	13065
Mediterranean E.	124	129	100	100	99	106	12321
Europe	99	96	99	100	99	103	132958

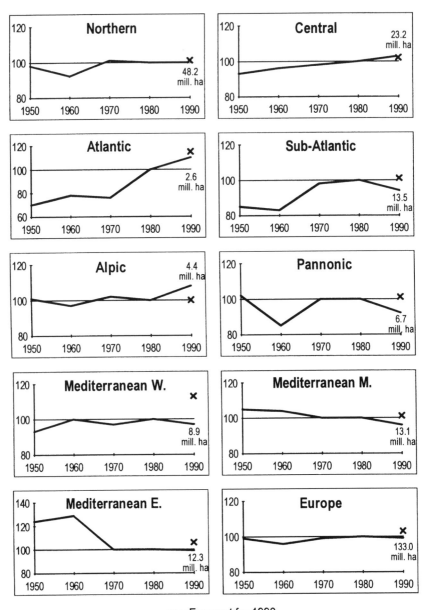

x = Forecast for 1990

Fig. 2.2   *Development of exploitable forest 1950-1990, by country groups. Area in 1980 equals 100. Forecast for 1990 made in 1980*

The area of Europe's exploitable forest, 133 million ha in 1990, has been more or less stable since 1950 in spite of the increasing total area of forest. This is because changing definitions of land classes have decreased the reported area of exploitable forests in Mediterranean Middle and East, especially in Turkey.

Increases have been proportionally greatest in Atlantic and Sub-Atlantic Europe, 5% and 11% respectively since 1950, consequent upon the successful afforestation of treeless lands.

The reclassification of certain exploitable forests as unexploitable has often concerned sites and stockings below average quality, especially in Mediterranean and Northern Europe. This has had the effect of increasing the quality of exploitable forests, overall improving the basis for developing management regimes and achieving multi-functional targets.

Area forecasts for 1990 have proved to be over-estimates in all country groups except Central and Alpic Europe. The increases in the latter regions, however, are not real but merely technical, caused by the first country-wide sampling inventories in Germany West and Switzerland.

The fact that the recorded development has proved less than that forecast is explained by two factors. Parts of the exploitable forests have been reclassified as unexploitable forest. A more important factor, though, is the fact that afforestation has been less active than was expected in 1980. The motives for afforestation and subsidising afforestation for wood production have weakened since the 1960s and 1970s.

2.2.2   Growing stock

Growing stock, GS, is the over-bark stem volume of living trees, in some cases including branches but excluding the tops of stems. The amount of growing stock in 1990, its development during the period 1950-1990, the coniferous percentage, C-%, as well as the 1980s forecast for 1990 of the growing stock by country groups, are presented in Table 2.6 and Fig. 2.3.

Europe's recorded growing stock, 18510 million m^3 in 1990, has increased by 43% since 1950 and by 16% since 1980. The relative increase has been greatest in Sub-Atlantic, Atlantic and Central Europe.

The growing stock has obviously been and is still under-estimated in those countries which have not accomplished statistically sound sampling inventories. Wherever the first sampling inventory has been carried out, the earlier estimates based on compiled results of management planning inventories or pure administrative guesses have proved to be under-estimates; e.g. the growing stock estimate increased by 20% in the 1960s in Austria and correspondingly by 100% and 15% in the 1980s in Germany West and in Switzerland.

It is impossible, however, to divide the recorded increases into the part caused by improved statistics and that caused by real increases in the growing stock.

The recorded volume of growing stock decreased in Atlantic and Pannonic Europe in the 1980s, although when calculated on the basis of recorded increment and fellings, the growing stock should have increased in both regions. Nevertheless, although there are many such incompatibilities in the statistics, the trend of the European growing stock has been for a continuous and significant increase since 1950.

Table 2.6   *Development of growing stock (index 1980 = 100) and its coniferous percentage in exploitable forests by country groups, 1950-1990 and forecast stock for 1990, made in 1980 (FC90)*

Country group	1950	1960	1970	1980	1990	FC90	1990
	Index: 1980 = 100						million m3
Northern	89	93	99	100	110	108	4721
Central	77	77	89	100	139	104	5099
Atlantic	51	56	72	100	99	130	233
Sub-Atlantic	56	57	86	100	116	107	1904
Alpic	78	78	87	100	121	106	1313
Pannonic	87	83	93	100	92	108	1431
Mediterranean W.	88	90	94	100	96	121	617
Mediterranean M.	79	76	78	100	109	106	1872
Mediterranean E.	106	108	114	100	120	102	1320
Europe	81	82	92	100	116	107	18510

Coniferous percentage					
Northern	83	83	84	84	84
Atlantic	42	49	50	58	59
Sub-Atlantic	42	47	38	40	39
Alpic	79	79	80	79	80
Pannonic	35	34	34	37	36
Mediterranean W.	66	66	57	62	62
Mediterranean M.	30	29	30	27	31
Mediterranean E.	61	61	57	55	58
Europe	64	65	63	63	64

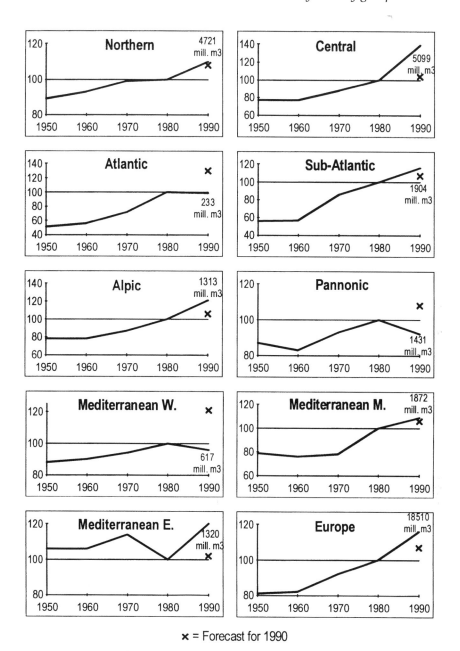

× = Forecast for 1990

Fig. 2.3   *Growing stock of exploitable forest, 1950-1990, and the forecast for 1990 made in 1980, by country groups (index 1980 = 100)*

Forecasts for 1990 made in 1980 have turned out to be under-estimates in all country groups except Atlantic, Pannonic and Mediterranean West Europe. For the whole of Europe, the increase of growing stock forecast was 7% while the recorded one is 16%. The difference is greatest in Central Europe, where the forecast was 4% and recorded increase 39% over the decade. This is mainly caused by the results of the first sampling inventory of Germany West.

The recorded proportion of coniferous forests in the total growing stock volume has been relatively stable in Europe and all country groups. It is greatest in Northern, Alpic and Central Europe, 84%, 80% and 71% respectively. It is smallest in Mediterranean Middle, Pannonic and Sub-Atlantic Europe, 31%, 36% and 39% respectively. The overall proportion is 64%.

2.2.3 Net annual increment

Net annual increment, NAI, is the gross annual increment minus natural losses. The estimates of it in 1950-1990, the coniferous percentage, C-%, and the forecast net annual increment for 1990 made in 1980 by country groups are presented in Table 2.7 and Fig. 2.4.

Comments concerning the reliability of recorded estimates are the same as those presented in the case of growing stock estimates. However, increment has been under-estimated even more than growing stock in the earlier Timber Trend Studies. Increment is more difficult to estimate than growing stock, even in sampling measurements. The theory of increment calculation on the basis of periodic measurements was not appropriately applied in Northern Europe in the 1960s and 1970s. In most countries, increment is estimated by yield tables representing fully dense stands. It is difficult to obtain unbiased results by this method for real forests where stands are of variable density and quality.

Europe's net annual increment, 584 million m³ over bark in 1990, has increased by 54% since 1950 and by 19% since 1980, which is proportionally more than the increase of growing stock. In the European total, the net annual increment of Germany West, which is missing in the 1990 Assessment, is estimated to be 7 m³/ha on the basis of increment rates reported by neighbouring countries.

The increases in NAI since 1980 are greatest in Central, Northern and Sub-Atlantic Europe, 36%, 22% and 22% respectively. It is smallest, 5%, in Atlantic Europe. In Northern Europe it is real within the limits of sampling error and possible systematic error in increment measurements. In some of those countries where the increment is estimated by methods other than sampling measurements, increment has increased by jumps during the last decade. It should have increased in Atlantic Europe more than that reported because of the large areas of planted forests. The earlier rising trend has stopped in Mediterranean West. Planted pine and eucalyptus stands have probably grown less than expected.

The 1980 increment value for Poland decreased by 18% from that estimated in 1970. One probable reason was the effect of atmospheric pollution. The 1990 value is greater than the 1980 value but smaller than the value in 1970. The increment reported by Czechoslovakia has increased compared with the 1980 Assessment.

Table 2.7   *Development of the net annual increment (index 1980 = 100) and its coniferous percentage in exploitable forest by country groups, 1950-1990, and the forecast increment for 1990, made in 1980 (FC90)*

Country group	1950	1960	1970	1980	1990	FC90	1990
			Index: 1980 = 100				1000 m3
Northern	89	92	92	100	122	108	178302
Central	81	89	102	100	136	104	132442
Atlantic	27	33	55	100	105	118	14382
Sub-Atlantic	66	69	76	100	122	103	73370
Alpic	65	70	92	100	114	105	27800
Pannonic	73	78	91	100	108	103	39825
Mediterranean W.	52	61	99	100	101	106	39036
Mediterranean M.	83	83	93	100	112	103	46130
Mediterranean E.	89	101	108	100	109	110	32630
Europe	77	82	93	100	119	106	583917
Coniferous percentage							
Northern	82	81	82	81	81		
Central	77	78	79	78	76		
Atlantic	70	75	82	81	82		
Sub-Atlantic	36	38	46	45	46		
Alpic	80	82	81	82	74		
Pannonic	28	31	30	31	31		
Mediterranean W.	64	69	64	71	67		
Mediterranean M.	23	26	23	28	28		
Mediterranean E.	45	45	50	53	54		
Europe	63	64	65	65	65		

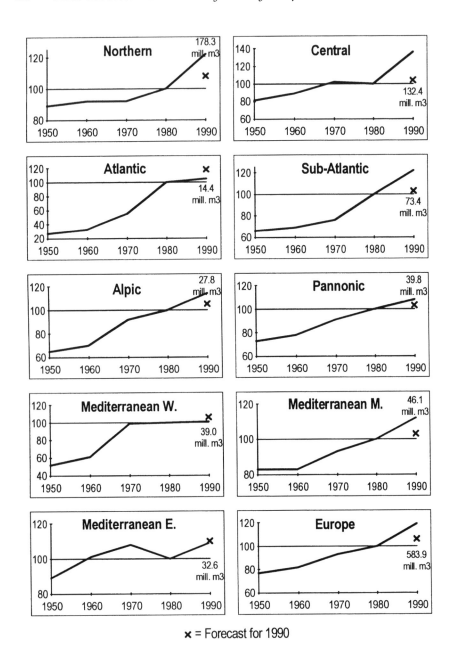

x = Forecast for 1990

Fig. 2.4   *Net annual increment of exploitable forest in 1950-1990 and the forecast for 1990, made in 1980, by country groups (index 1980 = 100)*

These examples illustrate a situation in which part of the increment estimates fluctuate from time to time, and they therefore poorly reflect the reality.

Austria, Finland and Sweden are the only countries where increment is estimated by regularly repeated sampling measurements. Their increment estimates have increased by 12%, 36% and 14% respectively during the 1980s. These increases were considerably greater than what was forecast in 1980 for the decade to 1990. This suggests that the fertilising effect of emissions, especially nitrogen and carbon dioxide, has increased rather than decreased the increment, at least temporarily.

The recorded 1990 increment exceeded that forecast in 1980 for 1990 in most country groups. For the whole of Europe, the forecast increase was 6% and the recorded one 19%.

The coniferous percentage in total increment has been relatively stable since 1950. Differences between country groups follow the differences of the coniferous percentages in growing stock. Conifers have increased their proportion, especially in Atlantic and Sub-Atlantic Europe due to large coniferous plantations there.

2.3 Total drain, harvest and increment in 1990

Total drain (D), natural losses (NL), fellings (F), logging residues (LR), removals over bark (Rob), bark (B), removals under bark (Rub), gross annual increment (GAI), and net annual increment (NAI), in 1990 by country groups are presented in Table 2.8. Relationships between these characteristics are presented by values of Northern Europe in Fig. 2.5.

The quality of these basic data is poor. For example, some countries have reported the same value for removals over and under bark. In some cases logging residues are missing. Consequently, fellings and drain are too small in relation to the removals under bark.

Europe's recorded annual drain from forest lands in 1990, 472 million m^3, is 175 million m^3 smaller than the recorded gross annual increment. In spite of the inaccuracies of both of these values, the difference indicates the annual accumulation of growing stock and the potential to increase the production of wood on a sustainable basis. The relationships between removals and increment are analysed in more detail below.

Table 2.8   **Drain, natural losses, fellings, logging residues, removals over bark, bark, removals under bark and gross and net annual increment on forest lands, by country groups, 1990**

Country group	D	NL	F	LR	Rob	B	Rub	GAI	NAI
					million m3				
Northern	137.20	6.76	130.44	9.46	120.98	14.16	106.83	192.40	185.64
Central	110.31	16.28	94.03	1.98	92.05	6.10	85.95	151.44	135.17
Atlantic	9.88	0.18	9.70	0.82	8.89	1.07	7.82	14.56	14.38
Sub-Atlantic	56.54	3.23	53.31	4.85	48.46	0.61	47.85	78.42	75.19
Alpic	23.42	0.26	23.16	1.25	21.91	1.91	20.00	30.30	30.04
Pannonic	28.73	3.71	25.02	1.26	23.76	2.19	21.57	45.29	41.58
Mediterranean W.	32.41	2.63	29.78	0.55	29.23	6.30	22.93	47.91	45.28
Mediterranean M.	32.89	0.48	32.41	3.55	28.87	4.47	24.40	48.62	48.14
Mediterranean E.	40.55	1.39	39.16	6.39	32.76	0.91	31.86	38.10	36.71
Europe	471.93	34.92	437.01	30.10	406.91	37.72	369.19	647.06	612.13

D, drain; NL, natural losses; F, fellings; LR, logging residues; Rob, removals over bark; B, bark; R, removals under bark; GAI, gross annual increment; NAI, net annual increment.

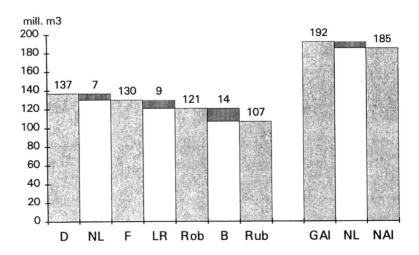

Fig. 2.5   *Drain, natural losses, fellings, logging residues, removals over bark, bark, removals under bark, gross and net annual increment on forest lands in Northern Europe, 1990 (see also Table 3.5)*

The recorded natural losses are 35 million m³, but reports are missing for six countries. Recorded natural losses are proportionally greatest in Poland, 10.0 million m³; in Czechoslovakia, 5.2 million m³; and in Spain, 2.5 million m³. The probable reasons for this are forest die-back in the former two countries, and wildfires in Spain.

Europe's natural losses in the 1980 Assessment equalled 32 million m³. Tree mortality and decay are probably increasing more than can be judged on the basis of the recorded values. The situation is analysed in Chapter 4 of this volume.

2.4   Removals 1950 to 1990 and fellings in 1990

The development of total annual removals, under bark, the coniferous percentage and removals forecast in 1980 for 1990 are presented in Table 2.9 and Fig. 2.6.

In principle, removal estimates should be more accurate than resource estimates such as growing stock and increment, especially in those countries where resources are not assessed by statistical sampling. There are, however, components in the total removals, such as fuel and household wood, which have been under-estimated, e.g. in countries where the proportion of the agricultural population is greater than average and where pasturing on forest lands is common.

Recorded removals between 1950 and 1990 increased by 25% for the whole of Europe. An exception to this is the Mediterranean Middle region, where removals declined from the high levels that prevailed after the Second World War. Removals also decreased in Pannonic Europe since 1970, probably because of the unstable political and economic situation. The increase was proportionally largest in Atlantic Europe, as well as in Mediterranean West and East. Large plantation forests have reached maturity or at least the thinning stage in the latter regions. The considerable increase of removals since 1960s recorded in Turkey may not be real. Removals of industrial wood have probably increased there, but a part of fuel and household wood may have been excluded in the earlier data.

Removals stabilized in Northern and Central Europe. In Northern Europe, the increasing consumption of industrial wood is covered by imports in spite of the abundant domestic resources.

Table 2.9   Development of total annual removals, under bark (index 1980 = 100), and its coniferous percentage and the forecast removals for 1990, made in 1980 (FC90)

Country group	1950	1960	1970	1980	1990	FC90	1990
		Index: 1980 = 100					1000 m3
Northern	83	91	106	100	101	106	106825
Central	81	77	83	100	103	99	85945
Atlantic	80	78	82	100	171	162	7816
Sub-Atlantic	79	84	100	100	114	118	47852
Alpic	67	82	86	100	107	120	19998
Pannonic	76	90	114	100	85	108	21570
Mediterranean W.	76	93	95	100	110	127	22925
Mediterranean M.	163	121	122	100	99	119	24398
Mediterranean E.	50	62	87	100	108	107	31856
Europe	83	86	98	100	104	109	369185
Coniferous percentage							
Northern	81	82	82	85	83		
Central	77	76	74	75	76		
Atlantic	30	43	61	69	86		
Sub-Atlantic	35	34	47	47	42		
Alpic	85	83	81	80	80		
Pannonic	29	29	27	29	26		
Mediterranean W.	43	44	51	63	54		
Mediterranean M.	23	18	22	27	22		
Mediterranean E.	20	39	50	54	57		
Europe	58	59	62	66	65		

In Alpic Europe, removals have increased more than average, by 67% from 1950 to 1990, probably because of the growth of the forest industry in Austria and the large availability of harvestable stock.

There is no strict positive correlation between the trends in removals and wood resources except in Atlantic Europe and Mediterranean West, where large plantations have increased the resources. In Northern and Central Europe, the national economies have lost, at least temporarily, the ability to utilize their wood resources fully.

For Europe as a whole, a 9% increase of removals by 1990 was forecast in 1980. The recorded increase was 4%. The recorded increase surpassed that forecast only in Central, Atlantic and Mediterranean East Europe. The real removals in 1990, however, were greater than recorded in the 1990 Assessment (cf. sections 2.5 and 2.6).

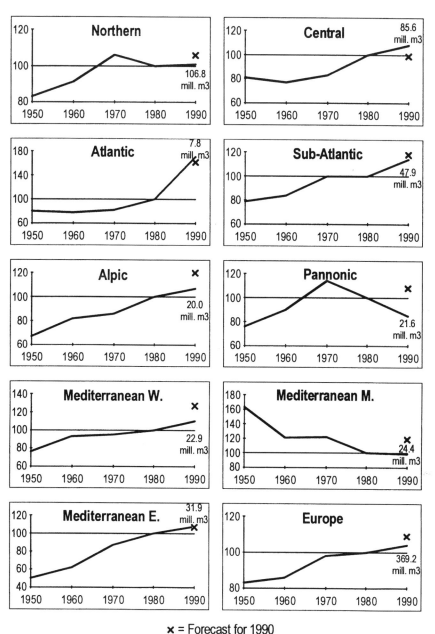

x = Forecast for 1990

Fig. 2.6   *Total removals under bark in 1950-1990 and 1980 forecast for 1990, by country groups (index 1980 = 100)*

The coniferous percentage of the removals has slightly increased in the totals, mainly due to the increasing removals from planted coniferous forests.

Removals, under bark, and fellings (total stemwood volume corresponding to the removals of exploitable forest in 1990) are presented and compared by country groups in Table 2.10. A probable range of removals divided by fellings, Rub/F, is 0.83 to 0.77. The greater the ratio, the smaller is the proportion of logging residues and bark from fellings. As the data of logging residues and bark are missing in some recordings, the Rub/F of 0.84 for the whole of Europe is higher than the actual value. Actual fellings, which would correspond to the figure of recorded removals, would be about 420 million m³ instead of 408 million m³.

Table 2.10   Removals under bark (Rub), fellings (F), and the proportion of removals of fellings (Rub/F) of exploitable forest in 1990 by country groups

Country group	Rub	F	Rub/F
	million m3		
Northern	102.7	125.2	0.82
Central	84.7	92.4	0.92
Atlantic	7.8	9.7	0.80
Sub-Atlantic	47.6	53.0	0.90
Alpic	19.5	22.6	0.86
Pannonic	19.1	22.0	0.87
Mediterranean W.	19.9	25.9	0.77
Mediterranean M.	23.7	31.5	0.75
Mediterranean E.	17.7	25.9	0.68
Europe	342.7	408.3	0.84

2.5 Wood-assortment structure of removals in the 1980s

The wood-assortment structure of removals by country groups in the 1980s is presented in Fig. 2.7.

In Northern Europe, removals hardly increased. The proportions of assortments were stable with the exception of a small increase in the proportion of coniferous sawlogs. Neither have the structure and total amount of removals changed much in Central Europe. Coniferous removals predominate. The proportions of coniferous sawlogs and fuelwood are greater than in Northern Europe.

The increasing trend of coniferous assortments is strong in Atlantic Europe, consequent upon afforestation stands reaching thinning or maturity stages. When compared with the structure of coniferous removals in Central Europe, the proportion of coniferous sawlogs is high in spite of the relatively short rotation ages. Obviously there is demand for the relatively small-sized and knotty sawlogs in domestic markets. Wood is used relatively little as fuel in Atlantic Europe.

Removals, especially coniferous, increased in Sub-Atlantic Europe. The increase of coniferous sawlogs is greater than that of small-sized timber. Afforestation stands are also coming into production. Broadleaved assortments account for little more than a half of the totals. Removals of fuelwood are considerable when compared with neighbouring regions.

Similarly, removals, especially coniferous ones, have increased significantly during the late 1980s in Alpic Europe, where the proportion of coniferous sawlogs is highest in Europe due to the applied thinning regimes and long rotation ages. The proportion of broadleaved industrial wood is small and that of fuelwood relatively large.

Removals reached their maximum in the years 1984 and 1985 in Pannonic Europe, after which they decreased sharply, possibly because of the disturbed economies in these countries. Broadleaved assortments predominate and the proportion of fuelwood is relatively large.

Removals are increasing significantly in Mediterranean West due to the large areas of planted fast-growing stands coming into production. Broadleaved assortments predominate, and the proportion of small-sized industrial wood is considerable, while that of fuelwood is relatively small. Removals of eucalyptus have increased and correspond to the amounts forecast earlier.

The amount of removals has levelled off in Mediterranean Middle. Broadleaved assortments predominate. The proportion of fuelwood is the second largest of the European regions.

Removals have dropped sharply in Mediterranean East in spite of the great need for wood there. A probable reason for this is the continued deterioration of forests in some of these countries. Broadleaved assortments comprise little more than a half of the total removals. The proportion of fuelwood is the largest of the European regions.

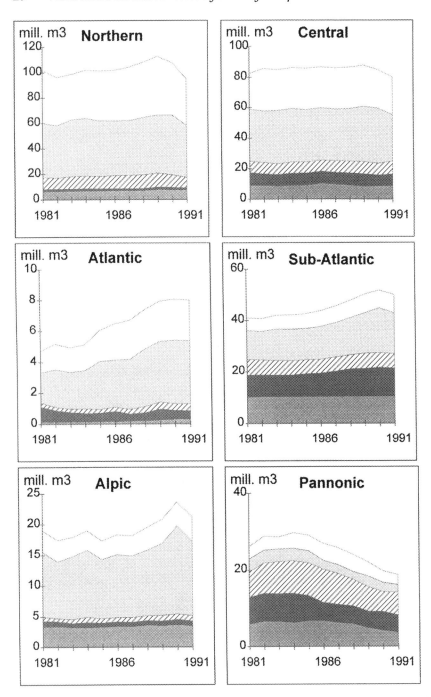

Fig. 2.7   *Removals by wood assortments in country groups and Europe in 1981-1990 (FAO Production statistics)*

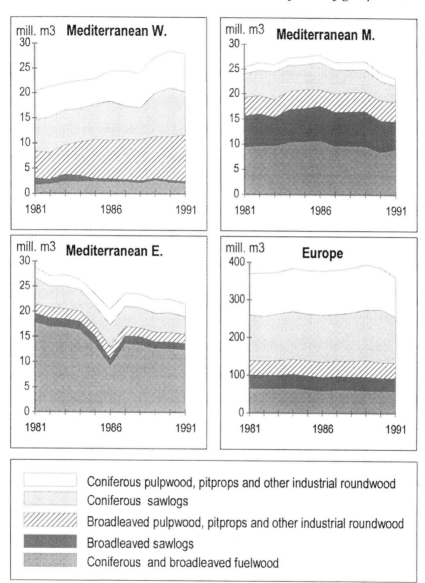

Fig. 2.7   Continued

The trend of removals in Europe during the 1980s was a steady although relatively small increase. Coniferous assortments, primarily sawlogs, made up the major part of this increase. Coniferous assortments accounted for about 67% of the total removals during the final years of the 1980s.

The assortment structure reflects the tree-species composition, age structure, management regimes and traditional uses of wood in each of the European country groups.

2.6   *Consistency of recorded removals*

Removals recorded in the latest Assessment concern 1990. The estimates, however, are not the means of the removals of the years 1989-1991 as in the earlier Assessments. They refer to different years in the different country reports during the late 1980s. By comparing them with the removals recorded in the FAO Yearbook of Forest Products 1990, the following observations can be made.

The consistency between the two information sources is relatively good in the cases of Northern, Atlantic, Sub-Atlantic and Mediterranean Middle Europe. There are marked differences between the figures in the case of other regions.

The differences are mostly caused by inaccuracies in reported statistics. Removals of the late years of the 1980s recorded in the FAO Yearbook are greater than the removals in the Assessment of 1990 because they include the increases caused by storm damage in Central and Alpic Europe. In the case of these two country groups, the estimates in the 1990 Assessment may better reflect the true development trends.

Total removals, under bark, in Europe are 369 million m^3 per annum in the 1990 Assessment and 414 million m^3 in the 1990 FAO Yearbook. By deducting the extra removals caused by storm damage, the true trend value around 1990 may be about 380 million m^3 which is 7% greater than the removals around 1980.

2.7   *The forest balance of exploitable forests*

Annual forest balance is a comparison between the gross annual increment and drain or net annual increment and fellings. In the periodic balance, the growing stock at the beginning and at the end of the balance period, as well as increment and drain during the period, are compared. The balance values and their symbols are as follows.

GS Growing stock volume of stem wood, over bark.

RGS Recorded growing stock as a result of sampling inventory or compiled from management planning statistics.

CGS Growing stock calculated for the end of a balance period on the basis of growing stock at the beginning of a period; and the difference between increment and drain during the period.

GAI Gross annual increment, total increase of stem wood, over bark, during a year. It includes the increment of those trees which are a part of growing stock at the end of the year and of those trees which have grown but died or been removed during the year.

D Drain, volume of those trees which are removed by forestry measures and natural causes from the growing stock.

NL Natural losses, volume of those trees which die of natural causes.

NAI Net annual increment, GAI minus NL.

F Fellings, D minus NL, volume of the trees which are removed from growing stock by forestry measures.

Annual balance is the difference either between GAI and annual drain, AD; or between NAI and annual fellings, AF.

CGS at the end of a balance period of 10 years can be calculated by one of the two formulas,

$$CGS + 10GAI \text{ minus } 10AD$$
$$CGS + 10NAI \text{ minus } 10AF$$

C/R, the ratio of calculated growing stock to recorded growing stock, is calculated by dividing CGS by RGS, both relating to the end of the period. A value of C/R between 1.05 and 0.95 demonstrates a good consistency between the balance values; and from 1.10 to 0.90 a satisfactory consistency. Good or satisfactory consistency does not always prove that the balance values equal the actual values in the forest, e.g. if both the growing stock and increment are under- or over-estimated proportionally by equal amounts, the consistency can appear to be good. Consequently, the main benefit of forest balances at the moment is to focus attention on the need to improve the accuracy of the balance characteristics.

Estimates of fellings in the 1990 Forest Resource Assessment are considered to represent the level of removals around 1990. The years for which countries have reported their removals vary, and the recorded fellings can also be taken as an average level of production during the period 1980-1990. In some cases the proportional values of the reported under- and over bark removals and fellings cannot be correct. All in all, the balances can be only approximate.

The increment balance for 1990 is presented for country groups in Table 2.11. For the whole of Europe, the net annual increment was 176 million m^3 or 43% higher than fellings. The increment is higher than fellings in all country groups. The excess is largest, 81%, in Pannonic and smallest, 23%, in Alpic Europe.

Table 2.11   Increment balance in 1990, million m^3

Country group	NAI	F	NAI minus F
Northern	178.3	125.2	+53.1
Central	132.4	92.4	+40.0
Atlantic	14.4	9.7	+4.7
Sub-Atlantic	73.4	53.0	+20.4
Alpic	27.8	22.6	+5.2
Pannonic	39.8	22.0	+17.8
Mediterranean W.	39.0	25.9	+13.1
Mediterranean M.	46.1	31.5	+14.6
Mediterranean E.	32.6	25.9	+6.7
Europe	583.0	408.1	+175.8

The periodic balance from 1980 to 1990, Table 2.12, is calculated by adding 10NAI to the growing stock in 1980 and deducting 10AF. The result is the calculated growing stock in 1990. The calculated growing stock, CGS, divided by the recorded stock, RGS, is C/R. It expresses the consistency between the balance values.

Table 2.12   Periodic balance in 1980-1990

Country group	RGS 1980	10NAI	10AF	CGS 1990	RGS 1990	C/R
			mill. m3			
Northern	4298	1783	1252	4829	4721	1.02
Central	3671	1324	924	4071	5099	0.80
Atlantic	235	137	97	275	233	1.18
Sub-Atlantic	1644	734	527	1852	1904	0.97
Alpic	1088	278	226	1140	1313	0.87
Pannonic	1553	398	221	1731	1431	1.21
Mediterranean W.	642	391	259	774	617	1.25
Mediterranean M.	1719	462	316	1865	1872	1.00
Mediterranean E.	1100	332	260	1172	1320	0.89
Europe	15950	5839	4082	17707	18510	0.96

According to the results, the consistency of balance values is best when the values are estimated by sampling inventories and drain and wood utilization studies. The overall consistency is, however, poor, although the country group differences even each other out to the average C/R for Europe as a whole.

Possible factors causing inconsistency are: under-estimated increment, inaccurately estimated growing stock and fellings not including correct logging residues, as well as bark and all wood used for fuel and household purposes. In countries where the proportion of agricultural population is high, all parts of grown wood may not be included in the balance values.

The balances for the 10-year periods between 1950 and 1990, Table 2.13, are calculated in the same way as the periodic balance 1980-1990. Increment is the value recorded at the end of each period. Fellings are means of the fellings at the beginning and end of each period.

The consistency of balances was apparently satisfactory in the 1950s. Only in Atlantic Europe was increment clearly under-estimated. Since 1950, consistency has decreased. This has been caused by the relatively large changes of individual balance values in sequential recordings.

Table 2.13   Development of approximate period balance of exploitable forest in 1950-1990 by country groups. C/R is the ratio of calculated (CGS) and recorded growing stock (RGS)

Country group		1950	1960	1970	1980	1990	
Northern	C/R		1.02	0.96	1.02	1.02	
Central	C/R		1.02	0.93	0.90	0.79	
Atlantic	C/R		0.33	0.93	1.10	1.31	
Sub-Atlantic	C/R		0.94	0.66	0.95	0.98	
Alpic	C/R		1.00	0.92	0.92	0.88	
Pannonic	C/R		1.09	0.92	0.97	1.19	
Mediterranean W.	C/R		1.01	1.20	1.17	1.27	
Mediterranean M.	C/R		0.92	0.94	0.82	1.00	
Mediterranean E.	C/R		1.07	0.99	1.14	0.87	
Europe	RGS	12994	13109	14691	15950	18510	Million m3
Europe	CGS		13274	13656	15457	17692	Million m3
Europe	C/R		1.01	0.93	0.97	0.96	Index

		1950-1960	1960-1970	1970-1980	1980-1990	
Europe	10NAI	4018	4535	4889	5839	Million m3
Europe	10AF	3738	3988	4123	4097	Million m3

10NAI = 10 years' net increment   10AF = 10 years' fellings

2.8   Increment estimates and climatic potentials, 1950-1990

Recorded estimates of gross annual increment per hectare of exploitable forest can be compared with the climatic potentials calculated by the CVP-index. The index was developed by Paterson in the early 1950s. It is based on the hypothesis that the increment of stem volume is primarily the function of climatic parameters in areas where the climate has had enough time to develop soils. Because of the differences in the soil parent materials, especially in regions which in geological terms have recently been under the effects of continental glaciation, the potential can only be applied with appropriate reservations for large forest areas.

   The mean temperature of the warmest month, the range between the mean temperature of the warmest and coldest month, the mean annual precipitation and the growing season in humid months are the independent parameters of the CVP-index. Paterson has given the climatic potentials for every country of the world in increment of stem volumes under bark, the values of which are converted into volumes over bark.

The CVP-index is based on the yield information available in the early 1950s. In critical analyses it should be remembered that climatic conditions now differ from those at the time the yield information stands were growing in their youngest stages. At least in Northern Europe, the climate was remarkably colder in the late 1800s and early 1920s than since the 1930s. Nitrogen precipitation and increased carbon dioxide in the atmosphere have also probably increased increment in more recent times. On the other hand, the yield of fully closed sample-plot stands is generally 10 to 15% greater than the actual potential of large forest areas. In the analysis of tree species composition, the age structure and density of the actual forests should be taken into consideration.

The gross annual increment per hectare recorded in the Assessments 1950 to 1990, as well as climatic potentials are presented in Table 2.14, Fig. 2.8 and Fig. 2.9 by country groups. Comments are presented in the country analyses in Chapter 3.

Table 2.14   *Recorded gross annual increments per hectare, 1950-1990, with respect to climatic potential by country groups*

Country group	Unit	1950	1960	1970	1980	1990	Potential
Northern	m3/ha	2.88	3.03	2.87	3.22	3.83	4.25
	%	68	71	68	76	90	100
Central	m3/ha	4.20	4.34	4.81	4.75	6.38	5.97
	%	70	73	81	80	107	100
Atlantic	m3/ha	2.29	2.51	4.18	5.96	5.57	8.90
	%	26	28	47	67	63	100
Sub-Atlantic	m3/ha	3.51	3.64	3.56	4.52	5.69	6.80
	%	52	54	52	66	84	100
Alpic	m3/ha	3.77	4.00	5.93	6.01	6.33	6.47
	%	58	62	92	93	98	100
Pannonic	m3/ha	4.60	4.36	4.94	5.32	6.11	5.04
	%	91	87	98	106	121	100
Mediterranean W.	m3/ha	1.33	1.68	3.27	4.57	4.68	4.83
	%	28	35	68	95	97	100
Mediterranean M.	m3/ha	2.23	2.21	2.98	3.23	3.76	6.84
	%	33	32	44	47	55	100
Mediterranean E.	m3/ha	1.81	1.99	2.47	2.66	2.92	5.87
	%	31	34	42	45	50	100
Europe	m3/ha	2.98	3.09	3.50	3.91	4.67	5.44
	%	55	57	64	72	86	100

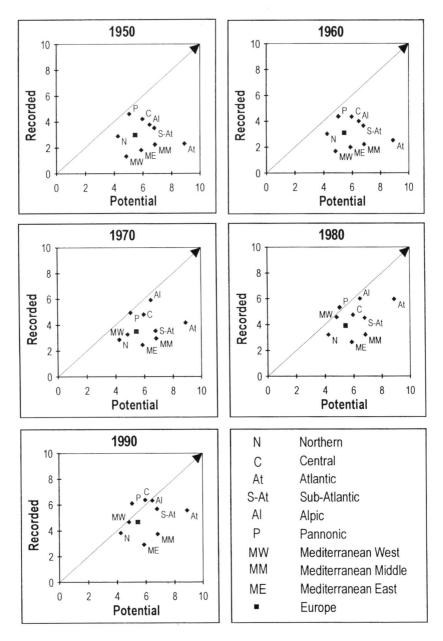

Fig. 2.8   *Recorded gross annual increment, 1950-1990, with respect to climatic potential, m³/ha, by country groups*

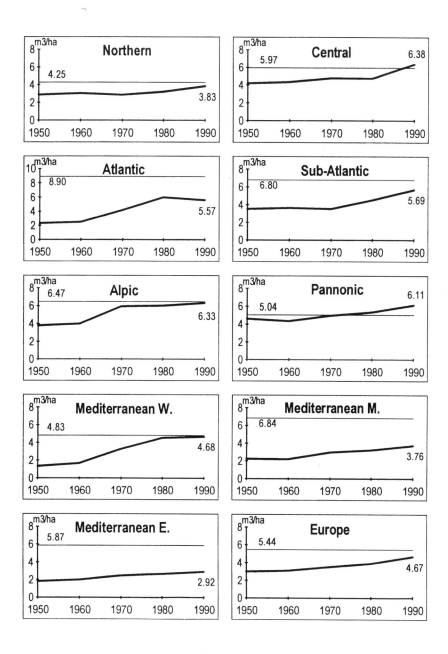

Fig. 2.9   *Recorded gross annual increment, 1950-1990, with respect to climatic potential, by country groups*

For the whole of Europe, recorded increment as a proportion of the potential has developed from 55% to 86% during the period 1950-1990. In Central, Alpic, Pannonic and Mediterranean West Europe the increment recorded in 1990 more or less equals the potential. In these country groups either the silvicultural quality of forests is good and stands fully dense, or there are many fast growing planted stands. In Atlantic Europe the increment recorded in 1990 is 63% of the potential. In this region, there are a lot of young planted stands at a stage where their increment is smaller than at later stages.

In Northern Europe, the increment recorded in 1990 is 90% of the potential. If the latest results of the National Forest Inventories of Finland and Sweden are taken into account, the measured increment of Northern Europe is approximately the same as the potential.

The figures are smallest in Mediterranean Middle and East, 54% and 50% respectively. In addition to the probable under-estimates, the density over large areas of these forests is lower than the full density.

If the climatic potentials had been used as a frame of reference in earlier times, the obvious under-estimates could have been, at least partly, eliminated by comparing the potentials with the recorded ones and country estimates with each other, recognizing the inconsistencies and systematically removing errors. Using this process the forecasts concerning the development of forest resources would have been more realistic than they were.

2.9   Forest ownership

Forest areas by the main ownership categories are presented in Table 2.15. It can be anticipated that the multi-functional benefits are generally provided best in public forests, and wood production in the forests owned by forest industry. Many of the poorly managed and neglected forests are owned by private people with small holdings. There are, however, public forests, in which working methods are at a relatively low technical level and their economy is unprofitable. The rotation ages in many public forests are so long that parts of forests are approaching stages at which deterioration sets in.

It should be relatively easy to develop management regimes which satisfy the multiple needs in public and industrial forests, as well as in large and medium sized forest holdings owned by private people.

Table 2.15   *Ownership of forest by countries and country groups in 1990*

Country or country group	Public		Industry	
	1000 ha	%	1000 ha	%
Finland	5275	26	1753	9
Norway	1392	16	346	4
Sweden	7038	29	5713	23
Denmark	127	27		
Germany W.	4092	54		
Germany E.	2044	70		
Poland	7201	83		
Czechoslovakia	4009	89		
Ireland	332	84		
United Kingdom	946	43		
Netherlands	159	48		
Belgium	273	44		
Luxembourg	38	45		
France	3514	27		
Austria	703	18		
Switzerland	766	68		
Hungary	1666	99		
Romania	6190	100		
Portugal	264	9	184	7
Spain	3250	39	350	4
Italy	2698	40		
Yugoslavia	5744	69		
Albania	1046	100		
Bulgaria	3386	100		
Greece	1946	77		
Turkey	8812	100		
Cyprus	127	91		
Israel	100	98		
Northern	13705	26	7812	15
Central	17473	72	0	0
Atlantic	1278	49	0	0
Sub-Atlantic	3984	28	0	0
Alpic	1469	29	0	0
Pannonic	7856	100	0	0
Mediterranean W.	3514	32	534	5
Mediterranean M.	9488	59	0	0
Mediterranean E.	14371	96	0	0
Europe	73138	49	8346	6

Table 2.15   Continued

Country or country group	Other private		Total		Average size of private holdings
	1000 ha	%	1000 ha	%	
Finland	13084	65	20112	100	small
Norway	6959	80	8697	100	medium
Sweden	11686	48	24437	100	medium
Denmark	339	73	466	100	small
Germany W.	3460	46	7552	100	medium
Germany E.	894	30	2938	100	
Poland	1471	17	8672	100	very small
Czechoslovakia	482	11	4491	100	
Ireland	62	16	394	100	
United Kingdom	1261	57	2207	100	
Netherlands	175	52	334	100	small
Belgium	347	56	620	100	small
Luxembourg	46	55	84	100	very small
France	9544	73	13058	100	very small
Austria	3174	82	3877	100	medium
Switzerland	364	32	1130	100	very small
Hungary	9	1	1675	100	
Romania			6190	100	
Portugal	2307	84	2755	100	small
Spain	4788	57	8388	100	very small
Italy	4052	60	6750	100	
Yugoslavia	2626	31	8370	100	
Albania			1046	100	
Bulgaria			3386	100	
Greece	567	23	2513	100	large
Turkey	44	0	8856	100	very small
Cyprus	13	9	140	100	
Israel	2	2	102	100	
Northern	31729	59	53246	100	
Central	6646	28	24119	100	
Atlantic	1323	51	2601	100	
Sub-Atlantic	10112	72	14096	100	
Alpic	3538	71	5007	100	
Pannonic	9	0	7865	100	
Mediterranean W.	7095	63	11143	100	
Mediterranean M.	6678	41	16166	100	
Mediterranean E.	626	4	14997	100	
Europe	67756	45	149240	100	

The ownership structure is least satisfactory in Poland, Sub-Atlantic Europe, Spain and Turkey, where the average size of private holdings is very small. Forest work can not be rationalised and mechanised properly on small holdings, as the income per owner is too low to maintain economic interests. Ownership is often a status symbol. The division of inherited holdings continues to fragment them into smaller and smaller lots.

The problem of small forest units is well known and much discussed, but difficult to solve. The shift of forest ownership from private to public in socialistic countries after World War II changed the problem of small forest lots into a problem of how to manage forests effectively. The heavy and expensive administration of forestry combined with low productivity were a burden in the socialist system. Relieving this burden and rearranging the ownership conditions restricts the application of effective management regimes and working methods in all the former East European socialist countries. The reorganization of forestry will obviously take more time than was expected some years ago.

2.10 Conclusions to Chapter 2

1.   The quality, consistency and reliability of recorded estimates have considerably improved since 1950. This positive development of forest resources culminated in the 1980 Assessment and retrogressed to some extent in the Assessment of 1990.

2.   Although the current information reflects better than before the actual forest resources, it is insufficient as a basis for policy implications. As long as many recorded estimates and their fluctuation from time to time are subjective random values rather than reliable statistics, there is no reliable basis for forest policy making.

3.   All European countries need to establish a system of repeated sampling inventories of the wood resources and multi-benefit characteristics of their forests. Removals and wood utilization studies should also be improved. Compared with the multiple values of forests and the environmental risks threatening them, the cost of reliable sampling inventories should not be an insurmountable economic obstacle.

4.   Reservation about, and even opposition towards, sampling inventories is often based on the implicit desire of forest administrators to have resources hidden; and in the case of forest owners to avoid unreasonable forest taxes. Where environmental, protective and amenity functions are valued higher than wood production, the undervaluing of the importance of sampling assessments can occur in spite of the fact that multiple functions cannot be managed properly without reliable quantitative resource information.

5.   The boundary between forest and other wooded land should be determined more accurately in Mediterranean Europe than it has been so far. The consistency of statistics on forests would be better if the potential site quality for growing trees were the principal - if not the only - criterion. An advantage of this criterion is the possibility to obtain more accurate estimates of lands available for afforestation. The boundary between exploitable and unexploitable forest land is also in need of clarification. Unambiguous land classes promote the development of appropriate management regimes.

6.   On the basis of the current forest lands statistics there are great possibilities for reforestation and stand improvements in Mediterranean Europe.

7.   The net increase of forest was at least 5 million ha between 1950 and 1990. The upward trend seems to level during the last decade. Forest lands have also increased statistically, but the development is confused by the changing definitions of land classes.

8.   Efficient and objective-oriented land-use policies and forest management measures are the means to increase forests and improve their quality. There is much evidence to show that a relatively extensive forest cover is possible in spite of a high population density.

9.   The coniferous percentage is considerable in Central, Atlantic, Sub-Atlantic and Alpic Europe compared with the deciduous dominant

tree-species composition under natural conditions. This decreases the stability of these forests compared with forests composed of native tree species.

10.   The motives both to afforest and to support afforestation seem to have decreased during the 1980s, at least for wood production.

11.   The growing stock of European forests increased by 43% between 1950 and 1990 and is still increasing. Removals have been less than both increment and the greatest sustainable yield. This has led to increasing age and density of stands and accumulation of mature stocks to levels which threaten ecological stability. Too small a demand for wood, especially for small-sized and low-quality assortments, and high logging costs in thinning, are typical factors in this development.

Exceptions to this picture occur in Mediterranean Europe where there are many stands of sub-optimal density.

The coniferous percentage in Europe as a whole has been relatively stable except in those countries where there are substantial areas of coniferous plantations.

12.   Increment has increased more than growing stock, partly because of earlier under-estimations of the former. In many countries there are fast growing coniferous and broadleaved forests which increase the total increment. Measurement of increment in some countries supports the conclusion that the fertilising effect of deposited emissions has increased rather than decreased the increment. Exceptions to this general picture are planted forests in Mediterranean West where the increment seems to have been smaller than that expected earlier.

13.   Natural losses, poorly known and recorded, are given as 32 and 35 million m^3 per year respectively in the Assessments of 1980 and 1990. The actual losses are probably greater than those recorded. They may be increasing because of the over-density and over-maturity of stands and because of the degrading effect of pollution. More efficient harvesting of natural losses could improve the self-sufficiency for wood in Europe.

14.   Removals have generally increased since 1950. The increase, however, has been slightly smaller than expected during the 1980s. The recorded fellings as removals under bark, are less than the actual fellings because of the missing estimates of bark and logging residues in some reports. Recorded fellings on exploitable forest are 408 million m^3 per annum, while actual fellings may be about 420 million m^3.

15.   The forest balance, despite inconsistencies between calculated and recorded values in many cases, demonstrates that there is great potential for increasing removals on the basis of sustainable yield. In the whole of Europe, the net annual increment of exploitable forests is 164 million m^3 greater than the recorded fellings in 1990. This is equivalent to approximately 134 million m^3 in removals under bark, and considerably greater than the net imports of wood to Europe.

 Percentages expressing the proportional increase of fellings required to equal the net increment in 1990 range from +26% in Mediterranean East to +81% in Pannonic Europe. The European average is +43%.

 In spite of the inaccuracy of the basic values, the unavoidable conclusion is that the European potential to increase removals on a sustainable basis is greater now than for many centuries.

16.   The recorded increment estimates have systematically moved closer to Paterson's climatic potentials over the past few decades. Europe's recorded increment was 55% of the potential in 1950 and was 86% in 1990. The recorded increment has recently exceeded the potential in some countries. Earlier under-estimates have been corrected, the silvicultural management of forests has improved and in some cases the age structure temporarily maintains increment at a higher level than possible in a forest with an even age-structure.

 The climatic conditions, silvicultural regimes and tree-species compositions have changed since Paterson's calculations were made. This calls for further studies to develop climatic yield indexes, which are of great value in monitoring forests in a changing environment.

17.   The proportional distribution of forest resources presented in Fig. 2.10 shows that Northern, Central and Sub-Atlantic Europe possess

the bulk of European forest and wood resources as well as removals. Climatic conditions for growing wood are best in Central, Atlantic, Sub-Atlantic, Alpic and Pannonic Europe. The cold climate in Northern Europe with evaporation greater than precipitation during the growing season, as well as the relatively poor quality of the growing stock in the Mediterranean region, retard increment and removals per unit area in these regions.

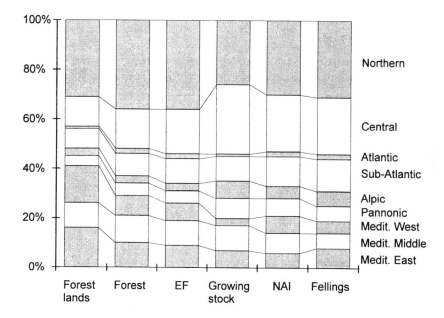

Fig. 2.10   *Proportional share of forest lands, forest, exploitable forest, growing stock, net annual increment and fellings of the European totals*

18.   The distribution of resource characteristics illustrates that increment and removals require, in spite of a cold climate, a much smaller growing stock per hectare in Northern Europe than in Central and Alpic Europe. In the latter regions, there is much low-profitability stock in the forests. The situation in Atlantic Europe is not clear because of the large proportion of young stands. Wood production there is obviously profitable if measured on the basis of growing-stock capital in the forest.

Land use in the Mediterranean groups is least effective, even given the unfavourable water regime found over large areas. Exceptions to this are the low-density forests, where the total production is composed of wood, cork, mast, etc. As a whole, there are great potentials to increase wood production in the Mediterranean area.

3 COUNTRY STATISTICS

Country statistics are based on the information reported to the Forest Resource Assessment 1990. More recently available resource information may differ from this information base in the case of some countries.

3.1 Northern Europe

Finland

Population 4.99 million, land area 30.46 million ha, forest lands 77%, forest 66%.

Resources per capita:

Land	6.10 ha
Forest	4.03 ha
Exploitable forest	3.91 ha
Net annual increment	14.38 m^3
Removals, under bark	9.18 m^3

The forest industries are Finland's major exporter. Since 1950, forests have been managed under efficient utilization, silvicultural and forest improvement regimes. Drainage to enhance growth affects about 5.7 million ha, 63% of the total wetland area.

Removals were greatest in the beginning of the 1960s and also high around 1970 and 1980. Since then removals have decreased. Increasing demand for industrial wood has been covered by imports which have been 5 to 6 million m^3 per annum.

The net annual increment exceeds fellings by 25%. Increment is rising due to the silvicultural and forest improvement inputs, and the gap between increment and fellings is increasing. Tree species and age class composition is good with respect to the high increment per hectare. The net increment has almost reached the climatic potential. Increment and fellings proportions of the growing stock volume are, respectively, 4.2% and 3.3%, demonstrating the high stock-capital intensity.

Topography is favourable for logging and silvicultural operations.

Unexploitable forests account for only 3% of the area of forest. All tree species are of local provenance and the biological stability of the forests is good. Natural losses, 2.2% of drain, are increasing because of the increasing areas of unthinned and over-mature stands. Health stability is satisfactory.

Public forests account for 26%, industrial forests 9%, and non-industrial private forests 65% of the area of forest. The average size of private forest holdings is relatively small. Demands concerning multi-benefit management and nature conservation, although growing, have been relatively small so far because of the large area of forest per capita.

The small size of private forest holdings and the unwillingness of owners to fully utilize their wood resources, together with the demands for nature conservation, restrict wood production. Demand for domestic industrial wood is smaller than potential supply, which results in the postponing of thinnings and regeneration of mature stands.

Norway

Population 4.24 million, land area 30.69 million ha, forest lands 31%, forest 28%.

Resources per capita:

Land	7.24 ha
Forest	2.05 ha
Exploitable forest	1.57 ha
Net annual increment	4.37 m^3
Removals, under bark	2.57 m^3

The Norwegian Forest Inventory comprises all types of forest lands except Finnmark, the northernmost county and the area above the coniferous forest limit.

Norway has long been an exporter of forest products. However, the forest sector's share of the annual gross domestic product has decreased since the 1950s.

From 1950 to 1990 the total standing volume of the productive forest has increased by 47%. The increase in the net annual increment was 44% over the same period. The recorded increment has been and remains smaller than the potential.

Removals decreased during the period from 1950-1970, but began to increase again in the 1980s, mainly because of efforts to intensify forestry.

This has increased the area of young stands under the age of 20 years.

Fellings have been markedly smaller than net increment. In the 1990 Assessment, the net annual increment was 49% greater than the annual fellings. Consequently, the proportion of mature and over-mature stands has increased, and the accumulation of mature stock will continue in the foreseeable future.

All tree species, except for some Sitka spruce planted on the west coast, are native. The genetic and ecological stability of the forests is good.

Norway is a mountainous country, creating difficult logging conditions. Unexploitable forests, 2.1 million ha, therefore account for 24% of the forest.

The proportion of small private forest holdings is the highest in Northern Europe, accounting for 77% of all forests and 85% of the exploitable forest recorded in the 1990 Assessment.

This ownership structure, together with difficult and expensive logging and transport conditions, are to be considered as the major constraints to obtain a more efficient production of wood in the future. The ownership structure is, however, an advantage for the environment as it entails a varied treatment of the forests.

Sweden

Population 8.56 million, land area 40.82 million ha, forest lands 69%, forest 60%.

Resources per capita:

Land	4.76 ha
Forest	2.85 ha
Exploitable forest	2.58 ha
Net annual increment	11.14 m^3
Removals, under bark	5.85 m^3

The relative importance of the forest sector and its exports in the national economy is the second highest in Europe, after Finland.

Removals were much smaller than the sustainable yield in the 1950s and 1960s. They reached a maximum around 1970 and have decreased since then. Recorded removals in 1990 were about 8% smaller than that forecast for 1990 made in 1980.

Because a part of the area of forest is located in the north where hauling distances are long and growing conditions poor, the unexploitable forest, 2.4 million ha, accounts for 10% of forest land.

Fellings have been smaller than increment since 1950. In the 1990 Assessment, the net annual increment is 58% greater than fellings. If the under-utilization continues, over-mature stands and their degradation will increase.

The increment recorded in 1990 equals the climatic potential and even exceeds it in the latest forest inventory estimate.

Tree species are native and the genetic stability of forests is good. Lodgepole pine is the only exotic grown in the productive forests, amounting to nearly 500 000 ha in 1990. The plantations are in the north and on mountain slopes. Because of fungal diseases, their stability is not yet secure in the harshest climate conditions.

The proportion of forest owned by private people, 50%, is the smallest, and the proportion of forests owned by industrial companies, 25%, is the largest in Northern Europe.

Environmental restrictions, the conservation of nature and the unwillingness of private forest owners to sell wood prevent the full utilization of wood resources. In the reshaped forest policy, as in many other countries, the target is to combine the production and environmental functions of forests.

Iceland

Population 0.25 million, land area 8.84 million ha, forest lands 1.5%, forest 0.1%.

Ancient forests, probably of birch, were destroyed and reduced to scrub land centuries ago. Pasturing has prevented trees from reoccupying forest lands.

A programme to rebuild soils, improve fields and pastures, and of afforestation began in 1974. Many tree species, including conifers, have been planted. Most of the stands have grown relatively well so far, but it is not yet sure that all stands will reach the closed forest stage.

The forest area in 1980 was 5000 ha, but it has grown to 11 000 ha in the 1990 Assessment. The current estimate of other wooded land is 123 000 ha. In the programme made in the 1970s, the afforestation target for the year 2000 was 100 000 ha. The principal function of forests is protective and environmental, although wood is used for fuel, fencing stakes, etc.

3.2   Central Europe

Denmark

Population 5.14 million, land area 4.25 million ha, forest lands and forest 11%.

Resources per capita:

Land	0.83 ha
Forest	0.09 ha
Exploitable forest	0.09 ha
Net annual increment	0.68 m³
Removals, under bark	0.38 m³

Denmark is a densely populated agricultural country. The forest area per capita is the smallest in Central Europe and one of the smallest in Europe as a whole. Land use is efficient. Most of the land outside other sectoral uses has been afforested.

Removals have increased relatively slowly but can be increased in the future when the proportionally large area of coniferous forest reaches the thinning and final cutting stages.

Beech is the dominant broadleaved tree. Its rotation age is fairly long because of the environmental and wood value of old, large-sized trees.

The net annual increment is 53% higher than fellings. The increment is relatively large per hectare because of the age structure in which young coniferous stands predominate. Stands of age 1 to 40 years account for 60% of forest and 67% of coniferous forests. Tree species composition is well balanced with regard to the wood production function and high landscape and recreational value of the forests. Increment, which exceeds the climatic potential, and the proportion of fellings in relation to growing stock are by far the highest in Central Europe. This demonstrates a very high stock-capital intensity.

Topography and infrastructure favour intensive silviculture and wood production. All the forests are exploitable with the exception of small areas for nature conservation.

Coniferous species have been re-introduced, because all the original coniferous forests were cut in historical times. Pine and spruce stands are in balance with the climatic and site conditions. There has, however, been considerable storm damage in spruce stands. The age and health stability is good, except in the spruce forests of western Denmark. Reported natural losses are 6.8% of drain, probably because of storm damage. Public ownership accounts for 27% of the forest and private individuals own 73%. The small average size of private forest holdings does not seem to restrict efficient forestry. Multi-benefit management works well and the conflicts between wood and non-wood benefits are smaller than in many other European countries.

Germany West (formerly Federal Republic of Germany)

Population 63.23 million, land area 24.41 million ha, forest lands 32% and forest 31%.

Resources per capita:

Land	0.39 ha
Forest	0.12 ha
Exploitable forest	0.12 ha
Net annual increment	0.80 m³
Removals, under bark	0.50 m³

The western part of Germany is densely populated and industrialized. The land use is efficient. The forest percentage is high considering the population density and efficient agriculture. All land outside other sectors is under forest. The forest industries are well developed and the production of wood-based panels and papers per capita is high.

Removals have been more or less constant since the 1950s and above the average only in years of great storm damage. The net annual increment was about 60% above average fellings at the end of the 1980s, while the potential for increasing sustainable removals is great. The current increment is higher than the climatic potential. The age-class structure is good, given the large increment per hectare. Tree-species composition favours multi-benefit management.

The annual net increment and fellings, 2.3% and 1.5% respectively, demonstrate a low capital intensity. The marginal rate of return on the considerable stock capital per hectare is negative.

Topography is generally favourable from the point of view of silvicultural and logging operations, but the costs of the current forestry methods result in low profitability. Unexploitable forest accounts for 2% of the forest and coppices 2% of the exploitable forest.

There may be a genetic risk in the large coniferous forests growing on the sites where oaks and beech are the natural climax species. This is symptomised by considerable storm damage. Reported natural losses are relatively small, but the high stand density and long rotations may cause them to increase in the future. Health stability is satisfactory, although the damage caused by air-borne pollutants is considered to be a serious threat to the forest in some areas.

Public ownership accounts for 52% and private individuals 46% of the forest. Non-wood benefits of forests are highly valued. A low industrial capacity restricts the full utilization of the wood resources.

Germany East (formerly German Democratic Republic)

Population 16.65 million, land area 10.52 million ha, forest lands 28% and forest 28%.

Resources per capita:

Land	0.63 ha
Forest	0.18 ha
Exploitable forest	0.15 ha
Net annual increment	1.11 m³
Removals, under bark	0.65 m³

The region has a long silvicultural tradition. Removals exceeded sustainability in the 1940s and 1950s, but the harvested areas were rapidly regenerated. The area of stands planted between 1950 and 1973 accounted for 48% of the exploitable forests. Consequently, young stands 1 to 40 years old account for 45% of the area of coniferous forests. The rotation age has been shorter than in the western part of Germany, also in the case of broadleaved forests.

The recorded increment exceeds the climatic potential by 11%. Because of the current age structure, the target of forest policy is to increase the volume of growing stock per hectare. The interest to utilize these wood resources has been relatively high, but insufficient demand for small-sized industrial wood has postponed thinnings especially in large pine forests. When the target growing stock has been reached, sustainably efficient production of wood will approach self-sufficiency.

The profitability of forestry has been poor, partly because of the costly administrative organization. A relatively great area, 16% of the area of forest, has been classified as unexploitable because of environmental functions and military restrictions. The rearrangement of forest ownership and forest administration, the reconstruction of forest industries and adopting a market economy will restrict the full utilization of wood resources.

Atmospheric pollution and climatic stresses have destroyed forests in limited areas, but these have been regenerated. About a quarter of all removals have been sanitary cutting in stands suffering from various types of damage.

Poland

Population 38.18 million, land area 30.44 million ha, forest lands 28%, forest 28%.

Resources per capita:

Land	0.80 ha
Forest	0.23 ha
Exploitable forest	0.22 ha
Net annual increment	0.82 m³
Removals, under bark	0.61 m³

The forest sector could be of much greater importance to the national economy than it is at present. In addition to maintaining self-sufficiency, Poland could be an exporter of forest products. The net annual increment is 12% greater than fellings, while actual increment is obviously greater than the reported one, which is 4.8 m³/ha compared with 6.8 and 8.1 m³/ha in East Germany and Czechoslovakia, respectively. The reported mean volume of growing stock, 163 m³/ha, may also be an under-estimation. The topography is favourable for effective forestry, and unexploitable forests account for 2% of the area of forest.

The genetic stability of the forests is satisfactory except in spruce stands on sites where broadleaved trees are the natural climax species. The reported natural losses are very high, 26% of the drain. This reflects the high density and age instability and also reveals a poor operational ability to carry out sanitary cuttings.

The forest is mostly owned by state, but private holdings of very small size account for 17% of the forest area. In the past they were under public control.

Multi-benefit management seems to be balanced, with the protective function reported to be of relatively high importance. Reorganization of the administration of forestry and the ownership structure restrict full utilization of the wood resources and the maintenance of environmental benefits.

Czechoslovakia (former)

Population 15.66 million, land area 12.54 million ha, forest lands 36%, forest 36%.

Resources per capita:

Land	0.80 ha
Forest	0.30 ha
Exploitable forest	0.29 ha
Net annual increment	1.98 m^3
Removals, under bark	1.16 m^3

The forest sector is of relatively great importance to the national economy. The full use of wood resources could maintain self-sufficiency and sustain exports of forest products. Forest industries are not yet sufficiently developed to fully consume the volume of removals produced.

The reported net increment is 53% greater than fellings, and it also exceeds climatic potential.

From the point of view of wood production, age-class and tree-species composition are both satisfactory. The actual rotation age and stand density exceed that required of an economic growing stock. Small increment and felling as percentages of growing stock demonstrate a low intensity of stock capital.

While the whole forest area is exploitable except some small nature conservation areas, the mountainous topography in parts of the country increases logging costs. Coppice makes up 2% of the exploitable forest area.

On low mountain slopes and plains there are extensive coniferous stands with genotypes originating outside of their provenance, despite which the stability of forests seems to be satisfactory. In spite of the damage caused locally by pollution, the reported increment of the growing stock exceeds that reported by neighbouring countries. Natural losses are high, 20% of drain, which demonstrates a poor ability to harvest trees in poor health because of over-density and over-maturity and/or damage caused by pollution and insects.

Forests have been owned and controlled by state organizations. Multi-benefit management seems to be well balanced. The protective function of the forests is reported to be of importance. Reorganization of the administration, economy and forest ownership structure restricts the full utilization of wood resources and even the maintenance of multi-benefits.

3.3 Atlantic Europe

Ireland

Population 3.50 million, land area 6.90 million ha, forest lands 6%, forest 6%.

Resources per capita:

Land	1.97 ha
Forest	0.11 ha
Exploitable forest	0.11 ha
Net annual increment	0.94 m³
Removals, under bark	0.40 m³

Fast growing planted coniferous forests are of increasing importance in the national economy of Ireland, where almost all the original forests were cleared in historical times for fields and pastures. The current increasing increment of the growing stock has already reached the volume of removals which equals Ireland's average self-sufficiency (taken as the average consumption of forest products in Europe).

Because the age structure is dominated by stands under 40 years age, increment is already approaching the climatic potential, and it exceeds fellings by 110%. Prescribed rotations demonstrate high marginal rates of return on growing stock capital.

The topography is favourable for wood production, and nearly all forests are exploitable. Most of the planted coniferous trees are exotics, in spite of which, the stability of forests has been good so far. Storm damage may increase with increasing stand age.

Public ownership accounts for 84% and private ownership 16% of the area of forest. The wood production function is reported to be the dominant aim, possibly because there is plenty of other land for recreation. Protective functions are relatively minor in the humid climate and low rolling terrain.

United Kingdom

Population 57.41 million, land area 24.09 million ha, forest lands 10%, forest 9%.

Resources per capita:

Land	0.42 ha
Forest	0.04 ha
Exploitable forest	0.04 ha
Net annual increment	0.19 m^3
Removals, under bark	0.11 m^3

Rapid increases of both removals from planted forests and production of the domestic processing industry are giving the forest sector increasing importance in the national economy as well as to its regions. Domestic production meets about 15% of total consumption. The aim of the large-scale afforestation projects, following World Wars I and II, was to create strategic wood reserves. This motive is now of low importance.

The reported net increment is 36% greater than fellings because of an age structure in which young stands predominate. The increment reported to the 1990 Assessment may be smaller than the actual one. Another factor causing the reported increment to be markedly smaller than the climatic potential is the relatively high proportion of broadleaved stands grown to the rotation age of about 120 years. These account for 28% of exploitable forest. The prescribed rotation ages indicate positive marginal rate of returns on the coniferous stock and negative rates on the broadleaved stock.

Efficient production of wood is not limited by topography and technically exploitable forests account for almost 100% of all woodland.

The stability of the forests is satisfactory, although on some terrain and soil types rotations have to be reduced because of wind-blow risk. A low level of natural losses demonstrate effective harvesting of windthrown trees. Public ownership accounts for 43% and private ownership 57% of the area of forest. The privatization of some state-owned forests has decreased public ownership in recent years. Multi-benefit management seems to be satisfactory and there are no environmental factors seriously restricting wood production, although the afforestation of upland grazing land is often restricted for environmental reasons.

3.4 Sub-Atlantic Europe

The Netherlands

Population 14.94 million, land area 3.99 million ha, forest lands 10%, forest 10%.

Resources per capita:

Land	0.23 ha
Forest	0.02 ha
Exploitable forest	0.02 ha
Net annual increment	0.16 m³
Removals, under bark	0.08 m³

The forest sector plays a minor role in the Netherlands' national economy, although the paper industry is of relatively great importance. In spite of a high population density, all land not in other uses has been intensively afforested. Stands are mostly young and many of them are composed of fast growing trees. Rotation ages are relatively short compared with those applied in Central Europe. The marginal rate of return on growing-stock capital is high.

The net annual increment exceeds fellings by 84%, largely because forest owners have under-estimated the cutting potentials and because of the high proportion of young and middle-aged stands. It is greater than the climatic potential. Removals are, at least potentially, increasing.

Topography and infrastructure both favour the efficient production of wood. Exploitable forests account for 99% of the area of forest and coppice 9% of exploitable forests.

From the point of view of multi-benefit forestry, tree-species composition is good and the stability of forests is satisfactory. Poplars grown in association with agricultural crops are gaining in importance. Public ownership accounts for 48% and private ownership 52% of the area of forest. Recreational benefits are of great importance. The demand for multi-benefits does not seem to restrict wood production.

Belgium

Population 9.84 million, land area 3.03 million ha, forest lands 20%, forest 20%.

Resources per capita:

Land	0.31 ha
Forest	0.06 ha
Exploitable forest	0.06 ha
Net annual increment	0.45 m^3
Removals, under bark	0.32 m^3

Forestry plays a minor role in the national economy, although the wood-based panels and paper industries, based partly on imported raw materials, have a relatively high per capita production. Land outside other uses has been effectively afforested.

Removals have increased markedly, because many of the young stands are reaching thinning and others final felling maturity. The net annual increment exceeds the climatic potential and is 43% greater than fellings. Tree-species and age-class compositions are favourable from the point of view of the high wood production potential per hectare. The average age of the current stands, including broadleaved stands, is about 60 years. The marginal rate of return of the growing stock is relatively high, but according to the prescribed rotation ages it may decrease in the future.

Topographic conditions favour the efficient production of wood. Logging conditions are also satisfactory in the hills of southern Belgium. Almost all forests are exploitable. The share of coppice, supplying a traditional demand for small-sized wood, accounts for 24% of exploitable forests.

The stability of forests seems to be good. Public ownership accounts for 44% and private ownership 56% of the area of forest. Wood production is of relatively high importance in spite of the need for recreational benefits in this densely populated country. Multi-benefit management works well.

Forests and forest management regimes differ greatly between the administrational regions of Belgium. These differencies cannot be reflected by the country averages.

Luxembourg

Population 0.37 million, land area 0.26 million ha, forest lands 34%, forest 33%.

Resources per capita:

Land	0.69 ha
Forest	0.23 ha
Exploitable forest	0.22 ha
Net annual increment	1.79 m^3
Removals, under bark	0.76 m^3

The role of forests is small in the national economy. Rotation ages of coniferous stands are around 80 years and of broadleaved stands 130 years or more. Broadleaved forests have an important environmental status. Public ownership amounts for 45% of forest and private ownership for 55%. The stability of the forests is good. Multiple benefit management is well balanced.

France

Population 56.44 million, land area 54.33 million ha, forest lands 26%, forest 23%.

Resources per capita:

Land	0.96 ha
Forest	0.23 ha
Exploitable forest	0.22 ha
Net annual increment	1.22 m^3
Removals, under bark	0.77 m^3

Forestry is of great importance to the national economy. The forest resources of France are the largest in the European Community with regard to forest area, and second largest with regard to growing stock and increment after the resources of the united Germany. If the species-assortment structure of forests were better, France would be self-sufficient with regard to forest products. Net increment, though smaller than the climatic potential, is nevertheless 37% higher than fellings.

Broadleaved stands account for 64% and coppice alone for 50% of exploitable forests. The average rotation age of broadleaved high forest is about 140 years. A large proportion of planted coniferous stands are of pines. Young age classes dominate and are in great need of thinnings and pruning. Spruce and fir forests on mountain slopes in eastern France are grown to long rotation ages. The general stability of forests is satisfactory. The reported natural losses, 6% of drain, are mostly caused by storm damage. Wild fires occur, mostly in Mediterranean scrub vegetation.

Public ownership accounts for 27% and private ownership for 73% of the area of forest. A great number of very small holdings restrict the development of private forestry. Another restriction is the large amount of low-quality broadleaved trees. Planted pines often have branchy stems and poor form, while the regeneration of coppice into high forests is expensive. Protective and recreational benefits have a high status which may also restrict wood production.

3.5 Alpic Europe

Austria

Population 7.71 million, land area 8.25 million ha, forest lands 47%, forest 47%

Resources per capita:

Land	1.07 ha
Forest	0.50 ha
Exploitable forest	0.43 ha
Net annual increment	3.11 m^3
Removals, under bark	1.96 m^3

Austria is an important net exporter of forest products, and export income per capita is the third highest in Europe after Finland and Sweden. The importance of the forest sector is reflected by management regimes guided by the results of the national forest inventory published at five-yearly intervals. Removals and wood resources have increased steadily since 1950, especially in the late 1980s.

The net increment, which is higher than the climatic potential, exceeds fellings by 27%. Tree species composition favours multiple benefit management. Young stands dominate the age structure of broadleaved forests. The actual rotation ages, however, are relatively long and the marginal rate of return on the growing stock is small. This is reflected by the low increment and fellings percentages of the growing stock, 2.3% and 1.8% respectively.

Topography, especially in the western Alpic regions, in addition to the heavy administration, increases the costs of forestry. Unexploitable forests account for 14% of the area of forest and coppice 3% of the exploitable forests.

In the lower mountain slopes and plains, there are spruce stands in localities where broadleaved trees are the original climax species; otherwise, the stability of the forests is good. Reported natural losses are exceptionally small.

Public ownership accounts for 18% and private ownership 82% of the area of forest. The average size of private holdings is higher than in Sub-Atlantic Europe, which is an advantage from the point of view of wood production. The status of both wood and the protective functions are high, and silviculture and hunting interests conflict. The tradition of maintaining a large growing stock per hectare and long rotation ages restricts the profitability of forestry.

Switzerland

Population 6.71 million, land area 3.97 million ha, forest lands 30%, forest 28%.

Resources per capita:

Land	0.59 ha
Forest	0.17 ha
Exploitable forest	0.16 ha
Net annual increment	0.90 m³
Removals, under bark	0.73 m³

Switzerland provides a good example of an efficient land-use policy. Forests have been well managed, even on the high Alpine slopes.

Removals have increased steadily since 1950, with the occasional fluctuations caused by storm damage. The net annual increment, probably an under-estimation, is smaller than the climatic potential. It exceeds fellings by at least 10%. Tree-species composition is good from the point of view of multi-benefit management.

The rotation age of coniferous stands is 80 to 250 years and that of broadleaved stands 80 to 180 years. Most of the oldest coniferous stands of exploitable forests are on the highest mountain slopes. The mean volume of the growing stock on exploitable forest, 329 m³/ha, is the greatest in Europe. Correspondingly, annual net increment and fellings are respectively 1.6% and 1.5% of growing stock. The marginal rate of return on the growing stock is negative because of the long rotation ages.

The mountainous topography restricts the profitability of wood production. Unexploitable forests account for only 3% of the area of forest and coppice 7% of the exploitable forests.

A part of the spruce stands grows outside the natural zone for this species, and the instability of forests caused by high density and over-maturity may be approaching the stage of biological degradation in the mountainous areas where the oldest stands are concentrated. Considering the efficient management and infrastructure, natural losses, 4% of drain, are relatively great. They are caused by avalanches and landslides in the mountains as well as by the inadequacy of forest roads and high logging costs. Degrading stands may be considered as contributing to the biodiversity.

Public ownership accounts for 68% and private ownership 32% of the area of forest. The small size of private forest holdings seems not to restrict wood production as much as in some other countries. The Federal and Cantonal Forest Administrations have promoted efforts to modernize forestry, but insufficient cooperation between forest owners hampers these efforts.

The different forest benefits are fairly well integrated. The apparent value of the growing stock restricts the multi-benefit management required to maintain a biologically sustainable forest structure needed to guarantee the desirable mix of functions.

3.6 Pannonic Europe

Hungary

Population 10.55 million, land area 9.21 million ha, forest lands 18%, forest 18%.

Resources per capita:

Land	0.87 ha
Forest	0.16 ha
Exploitable forest	0.13 ha
Net annual increment	0.94 m³
Removals, under bark	0.64 m³

Forestry plays a minor role in the national economy, but great efforts have been made to increase the self-sufficiency of wood by intensified silviculture and afforestation measures. Removals have increased markedly since 1950.

The net increment exceeds the climatic potential due to the fast growing new stands, and is 36% greater than fellings. Broadleaved stands make up 85% of growing stock, the highest proportion in Europe. Stands younger than 40 years account for 70% of the area of coniferous forest and 65% of that of broadleaved forest.

Prescribed rotation ages range from 60 to 70 years for pine, about 75 years for other conifers, 90 to 110 years for oak and 100 to 120 years for beech. They represent a much higher marginal rate of return on the growing stock than in Central Europe. Fast growing poplars and false acacia, managed on rotation ages of 15 to 30 years, have an important role in wood production.

Topography favours forest operations. Unexploitable forests account for 21% of the area of forest and coppice 29% of the exploitable forests.

The stability of forests seems to be satisfactory. Evaporation, which is greater than precipitation during growing season, restricts growth on the plains. Natural losses are relatively high, 12% of drain.

State ownership accounts for 99% of the area of forest. The wood and hunting functions are reported to have a higher status than protective and recreational functions. Reorganization of economic and social structures may impose limitations on the development of wood production.

Romania

Population 23.20 million, land area 22.95 million ha, forest lands 27%, forest 27%.

Resources per capita:

Land	0.99 ha
Forest	0.27 ha
Exploitable forest	0.23 ha
Net annual increment	1.37 m^3
Removals, under bark	0.64 m^3

The forest sector is of relatively great importance in the national economy. Removals increased markedly from 1950 to 1970 but have decreased since then. Romania could be self-sufficient in wood if its resources were fully utilized.

Net increment is greater than climatic potential and 98% higher than reported fellings, which may not include all the fuel and household wood consumed. Broadleaved trees account for 60% of the growing stock. Conifers are predominant in the mountains. Rotation ages are 100 to 140 years for both coniferous and broadleaved stands. The increment is 2.6% and fellings 1.4% of growing stock on which the marginal rate of return is very low.

Topography favours forest operations except in the highest mountains. The exploitable forest accounts for 87% of the area of forest, and coppice 8% of the exploitable forests. Over-maturity is the greatest threat to the stability of forests. Natural losses, 13% of drain, are relatively high and demonstrate a poor ability to harvest dying trees.

The state owns all the forests. The status of the wood production function is relatively high compared to other functions. Political and social conditions restrict the development of forest sector.

3.7 Mediterranean West

Portugal

Population 10.53 million, land area 8.66 million ha, forest lands 36%, forest 32%.

Resources per capita:

Land	0.82 ha
Forest	0.26 ha
Exploitable forest	0.22 ha
Net annual increment	1.12 m^3
Removals, under bark	0.76 m^3

The forest sector is of relatively great importance in the national economy, and also as a source of income in the developing regions where small-scale farming dominates. The removals and production of forest industries have increased markedly in the 1980s, while the production and exports of cork are the greatest in Europe.

Fellings are almost as great as net increment, which is much smaller than climatic potential. The relatively small average increment per hectare is consequent upon the low density and high age of the multi-benefit broadleaved forests.

Maritime pines account for 43% of the forest area, eucalyptus 8% and cork oak 22%. The increase of removals is mostly based on fast growing eucalyptus and pine plantations. Young stands dominate in the production forests proper.

Forest operations are difficult and expensive in mountainous regions where the wood production predominates. Unexploitable forests account for 15% of the area of forest and coppice 12% of the exploitable forests. Poorly utilized agricultural land comprises 3.6 million ha, of which at least half could be afforested. The stability of forests is satisfactory, but wild fires, affecting 48 000 ha a year on average, are a major threat to forests.

Private forests, mostly owned by farmers in small lots, account for 84% (the highest proportion in Europe) and industry forests 7% of the area of forest. There is great potential for developing silvicultural, logging and marketing conditions in private forests with state support. General conditions are conducive for practising multiple benefit management.

Spain

Population 38.96 million, land area 49.94 million ha, forest lands 51%, forest 17%.

Resources per capita:

Land	1.28 ha
Forest	0.22 ha
Exploitable forest	0.17 ha
Net annual increment	0.86 m^3
Removals, under bark	0.38 m^3

The forest sector plays a minor role in the national economy. Removals of industrial wood from fast growing plantation forests and an increasing industrial production base are of regional importance. Forests producing cork and mast in addition to household wood are essential parts of multi-benefit management, which includes agro-forestry.

Removals have increased markedly during the 1980s, especially from eucalyptus plantations. Large areas of planted pine forests are reaching the thinning stage. They were established mainly for protection against erosion and silting of hydro-electric reservoirs. Efforts to increase thinnings in these forests have not been successful because of mountainous conditions and insufficient financing. Forecasts of eucalyptus production have turned out to be over-optimistic.

The increment per hectare is relatively high because of the impact of the fast growing plantation forests. There is a great diversity of tree species. Unexploitable forests account for 22% of the area of forest and coppice 13% of the exploitable forests. There is a considerable area of other wooded land, 17.6 million ha, part of which should more correctly be classified as 'non-forest land available for afforestation'. There is much land available for afforestation, most of which is located in mountains and areas with poor growing conditions. Natural losses are relatively high, 2.5 million m³ per annum.

Public ownership accounts for 39%, industrial 4% and other private ownership 57% of the area of forest. Multi-benefit forestry predominates. Forestry activities have declined with the development of a new administrational framework in which there are 17 autonomous regions. Public attitudes do not favour wood production proper, partly because eucalyptus plantations are said to consume scarce water resources.

3.8 Mediterranean Middle

Italy

Population 57.66 million, land area 30.13 million ha, forest lands 28%, forest 22%.

Resources per capita:

Land	0.52 ha
Forest	0.22 ha
Exploitable forest	0.08 ha
Net annual increment	0.30 m³
Removals, under bark	0.13 m³

The forestry sector is of only minor importance in the national economy. Annual removals have decreased markedly and are now only half of what they were in 1950. There is considerable potential to increase removals again. Fellings are less than half of net increment, which is 57% of the climatic potential.

Broadleaved trees predominate and the species variety is great. Young age classes predominate, and a part of the stands are grown to a high rotation age. The best coniferous forests are located in northern Italy on the foothills of the Alps.

Unexploitable forests account for 35% of the area of forest and coppice 55% of exploitable forests. The area of other wooded land, 1.8 million ha, is relatively great. There are considerable possibilities for afforestation and to improve existing forest and other wooded land. The greater part of the coppice is deteriorating.

Public ownership accounts for 40% and private ownership 60% of the area of forest. Multi-benefit forestry is dominant. There is a considerable fire hazard with an average annual burnt over area of 50 000 ha.

National plans exist to develop forestry, but the results have so far been limited. Social conditions and general attitudes do not support wood production.

Yugoslavia (former)

Population 23.81 million, land area 25.54 million ha, forest lands 37%, forest 33%.

Resources per capita:

Land	1.07 ha
Forest	0.35 ha
Exploitable forest	0.33 ha
Net annual increment	1.24 m^3
Removals, under bark	0.64 m^3

The forest sector has had a relatively important role in the national economy and its role could be still greater if all forests were under efficient management.

Wood resources were heavily exploited after the World War II, since when removals have stabilized at a lower level, but they have collapsed during the 1990s, because of the internal conflicts.

The net increment has been markedly smaller than the climatic potential. This has mainly been caused by the high proportion of stands degraded by the harvesting of fuel and household wood, and by pasturing. Increment and fellings percentages of growing stock are at a low level, characteristic of the areas of Mediterranean Middle and East.

Broadleaved forests predominate with a wide tree-species composition. Unexploitable forests account for 7% of the area of forest, and coppice 38% of the exploitable forests.

There have been efforts to convert coppice into high forest but with little success. Difficult mountainous terrain has been an obstacle to the development of forestry. Old growing stock is accumulating, especially in beech stands which account for about 50% of all forest. An area of c. 50 000 ha of poplars has been planted.

Public ownership accounts for 69% and private ownership 31% of the area of forest. Multi-benefit management predominates. Protective and pasturing functions are relatively important. In the conditions of a disintegrating administrational and social framework, the quality of forestry decreases and the need for fuel and household wood will be great for the foreseeable future.

Albania

Population 3.25 million, land area 2.78 million ha, forest lands 52%, forest 38%.

Resources per capita:

Land	0.85 ha
Forest	0.32 ha
Exploitable forest	0.28 ha
Net annual increment	0.35 m³
Removals, under bark	0.55 m³

Because of the poor quality of forest statistics only some general observations can be made.

Forests have been used to supply the demands of consumption such as fuel and household wood, pasturing, as well as for small-scale industry. Removals have exceeded increment by a substantial margin. Forests have been degraded in spite of efforts to initiate improvement measures.

Unexploitable forests account for 13% of the area of forest, and coppice 39% of the exploitable forests. Broadleaved trees predominate. The mountainous terrain and changing social structure restrict the development of forestry. The state owns all forests.

Malta

Malta comprises a group of islands in the Mediterranean Sea with a population of 0.4 million and no forests proper.

3.9 *Mediterranean East*

Bulgaria

Population 9.01 million, land area 11.02 million ha, forest lands 33%, forest 31%.

Resources per capita:

Land	1.22 ha
Forest	0.38 ha
Exploitable forest	0.36 ha
Net annual increment	1.17 m³
Removals, under bark	0.39 m³

The forest sector has the potential to be of great importance in the national economy. The original forests, however, have been degraded over several centuries by heavy exploitation and pasturing, with the exception of small areas in inaccessible mountain areas. Removals have slightly decreased during the 1980s.

Unexploitable forests account for 5% of the area of forest, and coppice 27% of the exploitable forests. The tree-species composition, comprising broadleaved trees, sub-alpine conifers and Mediterranean trees, is poor from the point of view of the needs of the forest industries.

The reported increment, much smaller than the climatic potential, exceeds fellings by 87%. A part of fuel and household wood may not be included in the recorded fellings.

The state owns all forests. The social order is changing, which may slow down efforts to develop forestry.

Greece

Population 10.05 million, land area 12.93 million ha, forest lands 47%, forest 19%.

Resources per capita:

Land	1.29 ha
Forest	0.25 ha
Exploitable forest	0.23 ha
Net annual increment	0.36 m^3
Removals, under bark	0.25 m^3

Wood production plays a minor role in the national economy. Protective function against erosion and pasturing are of great importance. Removals have decreased since 1950 partly because of the continued degradation of the forests and partly because of the absence of roads in mountain forests. Removals have stabilized at current levels since the 1970s. Coniferous forests make up about 50% of growing stock and 28% of the removals. Most coniferous forests are located in mountain regions.

The reported fellings equal the net increment which is much smaller than the climatic potential. Unexploitable forests account for 9% of the area of forest, and coppice 48% of the exploitable forests. The area of other wooded land is 3.5 million ha, a great part of which could grow closed forest. The basic problem is wild fire as well as intensive pasturing. Degraded stands are being improved gradually by a forestry programme. Some areas where forest has been cleared for mining during the war will be placed under regular management.

Public ownership accounts for 77% and private ownership 23% of the area of forest. Uncontrolled harvesting of household wood is also practised in public forests.

Cyprus

Population 0.70 million, land area 0.92 million ha, forest lands 31%, forest 15%.

Resources per capita:

Land	1.31 ha
Forest	0.20 ha
Exploitable forest	0.13 ha
Net annual increment	0.08 m^3
Removals, under bark	0.10 m^3

Protection and landscape functions predominate in forest management. Conifers are the dominant species in forests proper. The best stands are planted while the natural stands are of low density. There are considerable areas of other wooded land.

Removals could be increased because a great part of stands are mature or over-mature. Wild fire frequency is great but fire control is efficient. Most of the forests are in public ownership.

Israel

Population 4.66 million, land area 2.03 million ha, forest lands 6%, forest 5%.

Resources per capita:

Land	0.44 ha
Forest	0.02 ha
Exploitable forest	0.02 ha
Net annual increment	0.06 m^3
Removals, under bark	0.02 m^3

The protection function predominates in forest management. Planted stands account for 64% of forest lands. There is a plan to afforest a further 50 000 ha. Most of the removals are harvested by thinnings and sanitation cuttings. Almost all of the forests are in public ownership. Fire frequency is relatively high.

Turkey

Population 58.69 million, land area 77.08 million ha, forest lands 26%, forest 11%.

Resources per capita:

Land	1.31 ha
Forest	0.15 ha
Exploitable forest	0.11 ha
Net annual increment	0.38 m^3
Removals, under bark	0.43 m^3

Potentially, the forest sector could be of much greater importance in the national economy. Forests have been heavily exploited by the rapidly growing rural population. Removals exceeded increment in the 1970s and early 1980s, but they decreased during the 1980s. All wood harvested for fuel and household purposes is probably not recorded.

The reported increment is about 50% of the climatic potential. Other wooded land, 11.3 million ha, growing scrub and isolated trees, accounts for 56% of forest lands. Unexploitable forests account for 25% of the area of forest and coppice 30% of the exploitable forests.

Mountainous and poorly accessible terrain is an obstacle to forestry operations. Coniferous trees make up more than half of the resources. A great number of tree species do not favour the production of industrial wood.

Virtually all forests are in public ownership. Traditional multi-benefit utilization dominates. There have been efforts to improve forests and develop the forest sector but the results have so far been modest.

Tables and figures

Table 3.1   *Forest lands by countries and their groups, 1990 (1000 ha)*
(for abbreviations, see p. iii)

Country or country group	Forest lands			Forest		
	Total	Forest	OWL	%	C-%	ha/cap
Finland	23373	20112	3261	66	92	4.03
Norway	9565	8697	868	28	67	2.05
Sweden	28015	24437	3578	60	94	2.85
Denmark	466	466		11	68	0.09
Germany W.	7754	7552	202	31	62	0.12
Germany E.	2981	2938	43	28	75	0.18
Poland	8672	8672		28	78	0.23
Czechoslovakia	4491	4491		36	64	0.29
Ireland	429	396	33	6	90	0.11
United Kingdom	2380	2207	173	9	72	0.04
Netherlands	334	334		10	60	0.02
Belgium	620	620		20	49	0.06
Luxembourg	87	85	2	33	36	0.23
France	14155	13110	1045	24	36	0.23
Austria	3877	3877		47	76	0.50
Switzerland	1186	1130	56	28	69	0.17
Hungary	1675	1675		18	16	0.16
Romania	6265	6190	75	27	31	0.26
Portugal	3102	2755	347	32	49	0.26
Spain	25622	8388	17234	17	47	0.22
Italy	8550	6750	1800	22	28	0.12
Yugoslavia	9453	8370	1083	33	22	0.35
Albania	1449	1046	403	38	17	0.32
Bulgaria	3683	3386	297	31	36	0.38
Greece	6032	2512	3520	19	38	0.25
Turkey	20199	8856	11343	11	58	0.15
Cyprus	280	140	140	15	99	0.20
Israel	124	102	22	5	57	0.02
Northern	60953	53246	7707	52	89	2.99
Central	24364	24119	245	29	70	0.17
Atlantic	2809	2603	206	8	75	0.04
Sub-Atlantic	15196	14149	1047	23	37	0.17
Alpic	5063	5007	56	40	74	0.35
Pannonic	7940	7865	75	25	28	0.23
Mediterranean W.	28724	11143	17581	19	47	0.23
Mediterranean M.	19452	16166	3286	28	24	0.13
Mediterranean E.	30318	14996	15322	14	50	0.18
Europe	194819	149294	45525	27	63	0.26

Table 3.1   *Continued*

Country or country group	EF		UEF	
	Area	%	Area	%
Finland	19511	97	601	3
Norway	6638	76	2059	24
Sweden	22048	90	2389	10
Denmark	466	100		0
Germany W.	7376	98	176	2
Germany E.	2476	84	462	16
Poland	8460	98	212	2
Czechoslovakia	4491	100		0
Ireland	394	99	2	1
United Kingdom	2207	100		0
Netherlands	331	99	3	1
Belgium	620	100		0
Luxembourg	82	96	3	4
France	12460	95	650	5
Austria	3330	86	547	14
Switzerland	1093	97	37	3
Hungary	1324	79	351	21
Romania	5413	87	777	13
Portugal	2346	85	409	15
Spain	6506	78	1882	22
Italy	4387	65	2363	35
Yugoslavia	7768	93	602	7
Albania	910	87	136	13
Bulgaria	3222	95	164	5
Greece	2289	91	223	9
Turkey	6642	75	2214	25
Cyprus	88	63	52	37
Israel	80	78	22	22
Northern	48197	91	5049	9
Central	23269	96	850	4
Atlantic	2601	100	2	0
Sub-Atlantic	13493	95	656	5
Alpic	4423	88	584	12
Pannonic	6737	86	1128	14
Mediterranean W.	8852	79	2291	21
Mediterranean M.	13065	81	3101	19
Mediterranean E.	12321	82	2675	18
Europe	132958	89	16336	11

Table 3.2   *Exploitable forests by countries and their groups, 1950–*
1990, and the 1980 forecast for 1990 (FC90)

Country or country group	1950	1960	1970	1980	1990
			Area (1000 ha)		
Finland	18200	18000	18734	19445	19511
Norway	6400	6450	6500	6600	6638
Sweden	22940	20009	23459	22230	22048
Denmark	438	374	373	365	466
Germany W.	6732	6838	6838	6960	7376
Germany E.	2749	2680	2680	2590	2476
Poland	7103	7524	8371	8410	8460
Czechoslovakia	3983	4119	3890	4185	4491
Ireland	124	171	268	347	394
United Kingdom	1535	1675	1521	2017	2207
Netherlands	250	260	256	294	331
Belgium	601	588	604	600	620
Luxembourg	81	81	81	81	82
France	11307	11000	13090	13340	12460
Austria	3139	2991	3230	3165	3330
Switzerland	982	982	972	935	1093
Hungary	1173	1214	1466	1596	1324
Romania	6326	5008	5861	5723	5413
Portugal	2467	3080	2848	2590	2346
Spain	6000	6000	5931	6506	6506
Italy	5648	6029	5342	3868	4387
Yugoslavia	7345	6833	7045	8500	7768
Albania	1330	1282	1240	1240	910
Bulgaria	2964	3169	3184	3300	3222
Greece	1847	1992	2312	2300	2289
Turkey	10330	10600	6642	6642	6642
Cyprus	171	147	173	100	88
Israel	14	66	43	66	80
Northern	47540	44459	48693	48275	48197
Central	21005	21535	22152	22510	23269
Atlantic	1659	1846	1789	2364	2601
Sub-Atlantic	12239	11929	14031	14315	13493
Alpic	4121	3973	4202	4100	4423
Pannonic	7499	6222	7327	7319	6737
Mediterranean W.	8467	9080	8779	9096	8852
Mediterranean M.	14323	14144	13627	13608	13065
Mediterranean E.	15326	15974	12354	12408	12321
Europe	132179	129162	132954	133995	132958

Table 3.2 **Continued**

Country or country group	1950	1960	1970	1980	1990	FC90
			Index: 1980 = 100			
Finland	94	93	96	100	100	103
Norway	97	98	98	100	101	100
Sweden	103	90	106	100	99	100
Denmark	120	102	102	100	128	100
Germany W.	97	98	98	100	106	100
Germany E.	106	103	103	100	96	101
Poland	84	89	100	100	101	104
Czechoslovakia	95	98	93	100	107	100
Ireland	36	49	77	100	114	113
United Kingdom	76	83	75	100	109	115
Netherlands	85	88	87	100	113	105
Belgium	100	98	101	100	103	102
Luxembourg	101	101	101	100	102	100
France	85	82	98	100	93	101
Austria	99	95	102	100	105	100
Switzerland	105	105	104	100	117	102
Hungary	73	76	92	100	83	104
Romania	111	88	102	100	95	100
Portugal	95	119	110	100	91	114
Spain	92	92	91	100	100	112
Italy	146	156	138	100	113	101
Yugoslavia	86	80	83	100	91	101
Albania	107	103	100	100	73	
Bulgaria	90	96	96	100	98	101
Greece	80	87	101	100	100	103
Turkey	156	160	100	100	100	109
Cyprus	171	147	173	100	88	100
Israel	21	100	65	100	121	108
Northern	98	92	101	100	100	101
Central	93	96	98	100	103	102
Atlantic	70	78	76	100	110	115
Sub-Atlantic	85	83	98	100	94	101
Alpic	101	97	102	100	108	100
Pannonic	102	85	100	100	92	101
Mediterranean W.	93	100	97	100	97	113
Mediterranean M.	105	104	100	100	96	101
Mediterranean E.	124	129	100	100	99	106
Europe	99	96	99	100	99	103

Table 3.3   **Growing stock of exploitable forests and the coniferous percentage (C-%) by countries and their groups, 1950-1990**

Country or country group	1950	1960	1970	1980	1990
	million m3				
Finland	1456	1430	1445	1568	1679
Norway	404	431	512	520	571
Sweden	1986	2134	2288	2210	2471
Denmark	43	45	45	46	54
Germany W.	990	960	1022	1100	2198
Germany E.	347	329	350	440	476
Poland	782	850	1049	1162	1380
Czechoslovakia	668	660	801	923	991
Ireland	7	9	15	32	30
United Kingdom	112	123	155	203	203
Netherlands	14	18	22	23	52
Belgium	46	57	70	73	90
Luxembourg	10	11	13	13	20
France	850	850	1307	1535	1742
Austria	578	567	681	776	953
Switzerland	273	280	270	312	360
Hungary	144	153	174	253	229
Romania	1200	1130	1268	1300	1202
Portugal	205	210	166	189	167
Spain	360	368	436	453	450
Italy	311	290	354	557	743
Yugoslavia	947	940	913	1082	1056
Albania	95	80	80	80	73
Bulgaria	248	248	264	298	405
Greece	105	107	150	159	149
Turkey	810	826	836	637	759
Cyprus	3	3	3	4	3
Israel			2	2	4
Northern	3846	3995	4245	4298	4721
Central	2830	2844	3267	3671	5099
Atlantic	119	132	170	235	233
Sub-Atlantic	920	936	1412	1644	1904
Alpic	851	847	951	1088	1313
Pannonic	1344	1283	1442	1553	1431
Mediterranean W.	565	578	602	642	617
Mediterranean M.	1353	1310	1347	1719	1872
Mediterranean E.	1166	1184	1255	1100	1320
Europe	12994	13109	14691	15950	18510

Table 3.3 **Continued**

Country or country group	1950	1960	1970	1980	1990
			C-%		
Finland	80	79	77	81	82
Norway	83	83	83	82	82
Sweden	85	85	86	85	85
Denmark	42	47	49	57	54
Germany W.	66	66	71	70	67
Germany E.	78	74	73	76	71
Poland	84	84	82	77	78
Czechoslovakia	66	74	75	74	73
Ireland	51	56	87	78	87
United Kingdom	41	49	46	55	55
Netherlands	86	83	73	65	56
Belgium	46	54	57	55	60
Luxembourg	22	20	15	17	18
France	41	46	37	39	38
Austria	84	84	85	84	83
Switzerland	69	69	69	67	73
Hungary	7	7	8	15	15
Romania	38	38	38	41	40
Portugal	57	57	51	61	66
Spain	71	71	59	62	61
Italy	34	34	36	36	36
Yugoslavia	29	28	28	22	29
Albania	30	30	30	30	21
Bulgaria	24	28	34	34	34
Greece	59	50	50	53	52
Turkey	72	72	66	65	72
Cyprus	100	100	100	100	100
Israel			40	50	50
Northern	83	83	83	83	84
Central	72	74	75	74	71
Atlantic	42	49	50	58	59
Sub-Atlantic	42	47	38	40	39
Alpic	79	79	80	79	80
Pannonic	35	34	34	37	36
Mediterranean W.	66	66	57	62	62
Mediterranean M.	30	29	30	27	31
Mediterranean E.	61	61	57	55	58
Europe	64	65	63	63	64

Table 3.4   *Net annual increment of exploitable forests and the coniferous percentage (C-%) by countries and their groups, 1950-1990*

Country or country group	1950	1960	1970	1980	1990
			1000 m3		
Finland	53300	52200	55800	61930	69664
Norway	14200	15600	15600	16980	17633
Sweden	62400	66300	63300	66940	91005
Denmark	2330	2440	2300	3405	3515
Germany W.	27950	31480	34000	34000	50894
Germany E.	13600	13600	13200	15000	16546
Poland	20100	23900	34600	28454	30464
Czechoslovakia	15500	15500	15700	16800	31023
Ireland	370	580	1900	2528	3294
United Kingdom	3310	4000	5700	11200	11088
Netherlands	1060	1060	1200	1241	2394
Belgium	2260	2430	2620	4500	4457
Luxembourg	170	266	400	330	664
France	36110	37780	41400	54000	65855
Austria	11210	12430	17290	19278	21980
Switzerland	4780	4780	5190	5200	5820
Hungary	4300	5200	6800	9612	8231
Romania	22800	23800	26900	27400	31594
Portugal	7460	8700	8200	11000	11286
Spain	12920	15120	30300	27830	27750
Italy	11220	11220	14000	11880	17475
Yugoslavia	20720	21040	22400	27294	27654
Albania	2200	2100	2100	2100	1001
Bulgaria	6400	6400	6500	6526	8870
Greece	3090	3790	4000	4100	3317
Turkey	17070	19980	21900	19210	20090
Cyprus	40	55	60	90	59
Israel			108	100	294
Northern	129900	134100	134700	145850	178302
Central	79480	86920	99800	97659	132442
Atlantic	3680	4580	7600	13728	14382
Sub-Atlantic	39600	41536	45620	60071	73370
Alpic	15990	17210	22480	24478	27800
Pannonic	27100	29000	33700	37012	39825
Mediterranean W.	20380	23820	38500	38830	39036
Mediterranean M.	34140	34360	38500	41274	46130
Mediterranean E.	26600	30225	32568	30026	32630
Europe	376870	401751	453468	488928	583917

Table 3.4   *Continued*

Country or country group	1950	1960	1970	1980	1990
			C-%		
Finland	76	76	79	78	78
Norway	87	82	80	78	80
Sweden	85	85	85	84	83
Denmark	61	64	65	74	73
Germany W.	74	70	76	79	77
Germany E.	84	84	84	77	76
Poland	89	88	84	82	80
Czechoslovakia	65	75	74	70	70
Ireland	87	82	84	97	100
United Kingdom	68	74	81	77	77
Netherlands	77	90	83	73	61
Belgium	60	63	58	54	66
Luxembourg	47	44	25	36	36
France	33	35	44	44	44
Austria	85	87	85	85	75
Switzerland	69	69	69	69	70
Hungary	11	7	7	15	17
Romania	31	36	36	37	35
Portugal	53	64	81	73	61
Spain	71	72	59	70	69
Italy	14	25	19	31	31
Yugoslavia	27	26	25	26	27
Albania	34	34	34	34	26
Bulgaria	20	25	31	33	33
Greece	59	50	45	46	47
Turkey	51	51	57	61	65
Cyprus	100	100	100	100	100
Israel			40	50	57
Northern	82	81	82	81	81
Central	77	78	79	78	76
Atlantic	70	75	82	81	82
Sub-Atlantic	36	38	46	45	46
Alpic	80	82	81	82	74
Pannonic	28	31	30	31	31
Mediterranean W.	64	69	64	71	67
Mediterranean M.	23	26	23	28	28
Mediterranean E.	45	45	50	53	54
Europe	63	64	65	65	65

Table 3.5   *Total drain, its parts and increment in 1990 by countries and their groups (1000 m³), ref. Fig. 2.5 (for abbreviations, see p. iii)*

Country or country group	D	NL	F	LR
Finland	58754	1294	57460	5413
Norway	14288	1523	12765	766
Sweden	64162	3944	60218	3280
Denmark	2720	185	2535	375
Germany W.	32557	738	31819	
Germany E.	11084	187	10897	
Poland	38545	9969	28576	1360
Czechoslovakia	25402	5200	20202	245
Ireland	1637	69	1568	
United Kingdom	8247	112	8135	815
Netherlands	1620	100	1520	
Belgium	3426		3426	
Luxembourg	360		360	47
France	51131	3131	48000	4800
Austria	17430	28	17402	821
Switzerland	5990	230	5760	430
Hungary	9470	1100	8370	813
Romania	19263	2614	16649	449
Portugal	11365	120	11245	373
Spain	21042	2512	18530	173
Italy	8254	294	7960	338
Yugoslavia	22581	184	22397	3208
Albania	2056		2056	
Bulgaria	5095	340	4755	837
Greece	4426	1050	3376	514
Turkey	30224		30224	5035
Cyprus	720		720	
Israel	81		81	8
Northern	137204	6761	130443	9459
Central	110308	16279	94029	1980
Atlantic	9884	181	9703	815
Sub-Atlantic	56537	3231	53306	4847
Alpic	23420	258	23162	1251
Pannonic	28733	3714	25019	1262
Mediterranean W.	32407	2632	29775	546
Mediterranean M.	32891	478	32413	3546
Mediterranean E.	40546	1390	39156	6394
Europe	471930	34924	437006	30100

Table 3.5   *Continued*

Country or country group	Rob	B	Rub	GAI	NAI
Finland	52047	6235	45812	73029	71735
Norway	11999	1091	10908	20069	18546
Sweden	56938	6833	50105	99301	95357
Denmark	2160	215	1945	3700	3515
Germany W.	31819		31819	51632	50894
Germany E.	10897		10897	18705	18518
Poland	27216	4074	23142	41184	31215
Czechoslovakia	19957	1815	18142	36223	31023
Ireland	1568	157	1411	3363	3294
United Kingdom	7320	915	6405	11200	11088
Netherlands	1520	269	1251	2519	2419
Belgium	3426	307	3119	4457	4457
Luxembourg	313	31	282	664	664
France	43200		43200	70780	67649
Austria	16581	1483	15098	24000	23972
Switzerland	5330	430	4900	6300	6070
Hungary	7557	837	6720	11002	9902
Romania	16200	1350	14850	34291	31677
Portugal	10872	2857	8015	11913	11793
Spain	18357	3447	14910	36000	33488
Italy	7622	366	7256	17769	17475
Yugoslavia	19189	3836	15353	29727	29543
Albania	2056	267	1789	1123	1123
Bulgaria	3918	392	3526	10917	10577
Greece	2862	366	2496	4699	3649
Turkey	25189		25189	22135	22135
Cyprus	720	140	580	59	59
Israel	73	8	65	294	294
Northern	120984	14159	106825	192399	185638
Central	92049	6104	85945	151444	135165
Atlantic	8888	1072	7816	14563	14382
Sub-Atlantic	48459	607	47852	78420	75189
Alpic	21911	1913	19998	30300	30042
Pannonic	23757	2187	21570	45293	41579
Mediterranean W.	29229	6304	22925	47913	45281
Mediterranean M.	28867	4469	24398	48619	48141
Mediterranean E.	32762	906	31856	38104	36714
Europe	406906	37721	369185	647055	612131

Table 3.6   **Annual removals and the coniferous percentage (C-%) by countries and their groups, 1950-1990**

Country or country group	1950	1960	1970	1980	1990
			1000 m3		
Finland	39367	44100	44080	45642	45812
Norway	10327	9100	8667	8945	10908
Sweden	37833	42500	59051	50846	50105
Denmark	1973	1780	2230	2109	1945
Germany W.	28329	25710	27801	31461	31819
Germany E.	13043	7500	7169	9990	10897
Poland	13146	16200	18289	20721	23142
Czechoslovakia	11362	13000	13501	18803	18142
Ireland	169	350	346	379	1411
United Kingdom	3468	3220	3432	4191	6405
Netherlands	600	730	905	896	1251
Belgium	2165	2358	2561	2251	3119
Luxembourg	186	183	218	292	282
France	30421	31950	38190	38581	43200
Austria	9102	11920	12070	14343	15098
Switzerland	3492	3540	4005	4398	4900
Hungary	2472	3500	4998	6241	6720
Romania	16654	19300	23776	19015	14850
Portugal	4517	5830	6247	8420	8015
Spain	11250	13500	13666	12376	14910
Italy	14052	11950	10885	8741	7256
Yugoslavia	25070	16380	16963	13622	15353
Albania	1100	1600	2330	2330	1789
Bulgaria	5062	5400	5050	4459	3526
Greece	3255	3030	2868	2590	2496
Turkey	6556	9800	17823	22376	25189
Cyprus	38	46	46	84	580
Israel	8	29	81	118	65
Northern	87527	95700	111798	105433	106825
Central	67853	64190	68990	83084	85945
Atlantic	3637	3570	3778	4570	7816
Sub-Atlantic	33372	35221	41874	42020	47852
Alpic	12594	15460	16075	18741	19998
Pannonic	19126	22800	28774	25256	21570
Mediterranean W.	15767	19330	19913	20796	22925
Mediterranean M.	40222	29930	30178	24693	24398
Mediterranean E.	14919	18305	25868	29627	31856
Europe	295017	304506	347248	354220	369185

Table 3.6 **Continued**

Country or country group	1950	1960	1970	1980	1990
			C-%		
Finland	73	76	72	82	80
Norway	87	88	90	92	92
Sweden	87	87	88	86	83
Denmark	46	50	57	59	58
Germany W.	70	67	67	69	77
Germany E.	87	84	83	79	78
Poland	91	86	82	81	75
Czechoslovakia	76	79	75	76	78
Ireland	32	89	88	92	98
United Kingdom	30	38	59	67	83
Netherlands	51	73	66	67	54
Belgium	54	53	63	65	66
Luxembourg	53	36	37	38	51
France	33	32	45	46	40
Austria	88	86	85	84	82
Switzerland	78	73	70	67	74
Hungary	2	3	5	6	7
Romania	33	34	32	36	35
Portugal	58	67	72	66	59
Spain	37	34	42	62	51
Italy	18	10	12	17	21
Yugoslavia	25	23	27	33	24
Albania	34	34	34	34	13
Bulgaria	29	24	30	27	29
Greece	19	7	24	26	28
Turkey	14	57	60	63	63
Cyprus	66	91	91	98	94
Israel	50	50	50	53	74
Northern	81	82	82	85	83
Central	77	76	74	75	76
Atlantic	30	43	61	69	86
Sub-Atlantic	35	34	47	47	42
Alpic	85	83	81	80	80
Pannonic	29	29	27	29	26
Mediterranean W.	43	44	51	63	54
Mediterranean M.	23	18	22	27	22
Mediterranean E.	20	39	50	54	57
Europe	58	59	62	66	65

Table 3.7   Annual fellings on exploitable forests by countries and their groups, 1950-1990

Country or country group	1950	1960	1970	1980	1990
			1000 m3		
Finland	48850	54720	54700	55690	55857
Norway	11882	10567	10240	10894	11814
Sweden	47150	53020	72220	63234	57543
Denmark	2180	1970	2470	2100	2285
Germany W.	35225	31344	33689	37164	31819
Germany E.	15775	9070	9360	12040	10833
Poland	16030	19710	22140	25300	27318
Czechoslovakia	12620	14490	15000	21460	20152
Ireland	190	390	390	650	1568
United Kingdom	4245	3930	4230	4580	8135
Netherlands	780	950	1180	1180	1300
Belgium	2660	2910	3150	2775	3326
Luxembourg	225	215	255	348	360
France	38830	44190	38170	41460	48000
Austria	11220	14900	15090	15210	17272
Switzerland	4165	4220	4795	4800	5300
Hungary	3000	3810	5950	7540	6057
Romania	18700	21600	26700	20700	15950
Portugal	5355	6940	7455	10800	10879
Spain	14200	16990	16810	13340	15014
Italy	15405	20800	11905	8060	7960
Yugoslavia	35150	22960	23790	19940	21957
Albania	1540	2240	3262	3262	1629
Bulgaria	6975	7425	7020	6030	4755
Greece	3970	3670	3520	2845	3367
Turkey	7240	10940	19900	19570	17152
Cyprus	48	58	46	70	577
Israel	10	40	104	90	69
Northern	107882	118307	137160	129818	125214
Central	81830	76584	82659	98064	92407
Atlantic	4435	4320	4620	5230	9703
Sub-Atlantic	42495	48265	42755	45763	52986
Alpic	15385	19120	19885	20010	22572
Pannonic	21700	25410	32650	28240	22007
Mediterranean W.	19555	23930	24265	24140	25893
Mediterranean M.	52095	46000	38957	31262	31546
Mediterranean E.	18243	22133	30590	28605	25929
Europe	363620	384069	413541	411132	408257

Table 3.8 *Forest balance: Growing stock in 1980 (RGS 1980) + 10 years' net increment (10NAI) minus 10 years' fellings (10AF) = calculated growing stock in 1990 (CGS 1990); recorded growing stock in 1990 (RGS 1990) (ref. section 2.7)*

Country or country group	RGS 1980	10NAI	10AF	CGS 1990	RGS 1990	C/R
			1000 m3			
Finland	1568	697	559	1706	1679	1.02
Norway	520	176	118	578	571	1.01
Sweden	2210	910	575	2545	2471	1.03
Denmark	46	35	23	58	54	1.08
Germany W.	1100	509	318	1291	2198	0.59
Germany E.	440	165	108	497	476	1.04
Poland	1162	305	273	1194	1380	0.87
Czechoslovakia	923	310	202	1031	991	1.04
Ireland	32	33	16	49	30	1.64
United Kingdom	203	111	81	233	203	1.15
Netherlands	23	24	13	34	52	0.65
Belgium	73	45	33	84	90	0.94
Luxembourg	13	7	0	19	20	0.97
France	1535	659	480	1714	1742	0.98
Austria	776	220	173	823	953	0.86
Switzerland	312	58	53	317	360	0.88
Hungary	253	82	61	275	229	1.20
Romania	1300	316	160	1456	1202	1.21
Portugal	189	113	109	193	167	1.16
Spain	453	278	150	581	450	1.29
Italy	557	175	80	652	743	0.88
Yugoslavia	1082	277	220	1139	1056	1.08
Albania	80	10	16	74	73	1.01
Bulgaria	298	89	48	339	405	0.84
Greece	159	33	34	158	149	1.06
Turkey	637	201	172	666	759	0.88
Cyprus	4	1	6		3	0.00
Israel	2	3	1	4	4	1.06
Northern	4298	1783	1252	4829	4721	1.02
Central	3671	1324	924	4071	5099	0.80
Atlantic	235	144	97	282	233	1.21
Sub-Atlantic	1644	734	527	1852	1904	0.97
Alpic	1088	278	226	1140	1313	0.87
Pannonic	1553	398	221	1731	1431	1.21
Mediterranean W.	642	391	259	774	617	1.25
Mediterranean M.	1719	462	316	1865	1872	1.00
Mediterranean E.	1100	326	260	1168	1320	0.88
Europe	15950	5841	4081	17711	18510	0.96

Table 3.9   **Gross annual increment (GAI) of exploitable forests, climatic potential (CP) and increment as percent of growing stock by countries and their groups in 1990**

Country or country group	Area	GAI		CP	
	ha	m3	m3/ha	m3	m3/ha
Finland	19511	70849	3.63	74142	3.80
Norway	6638	18928	2.85	35181	5.30
Sweden	22048	94769	4.30	94806	4.30
Denmark	466	3700	7.94	2423	5.20
Germany W.	7376	51632	7.00	44994	6.10
Germany E.	2476	16713	6.75	15104	6.10
Poland	8460	40183	4.75	49068	5.80
Czechoslovakia	4491	36223	8.07	27395	6.10
Ireland	394	3294	8.36	3940	10.00
United Kingdom	2207	11200	5.07	19201	8.70
Netherlands	331	2499	7.55	2052	6.20
Belgium	620	4683	7.55	4464	7.20
Luxembourg	82	698	8.51	508	6.20
France	12460	68901	5.53	84728	6.80
Austria	3330	22000	6.61	20646	6.20
Switzerland	1093	6010	5.50	7979	7.30
Hungary	1324	9145	6.91	7944	6.00
Romania	5413	32026	5.92	25982	4.80
Portugal	2346	11400	4.86	17360	7.40
Spain	6506	30000	4.61	25373	3.90
Italy	4387	17769	4.05	31148	7.10
Yugoslavia	7768	27878	3.59	52046	6.70
Albania					
Bulgaria	3222	10368	3.22	16754	5.20
Greece	2289	4278	1.87	12132	5.30
Turkey	6642	20791	3.13	42509	6.40
Cyprus					
Israel					
Northern	48197	184546	3.83	204129	4.24
Central	23269	148451	6.38	138984	5.97
Atlantic	2601	14494	5.57	23141	8.90
Sub-Atlantic	13493	76781	5.69	91752	6.80
Alpic	4423	28010	6.33	28625	6.47
Pannonic	6737	41171	6.11	33926	5.04
Mediterranean W.	8852	41400	4.68	42733	4.83
Mediterranean M.	12155	45647	3.76	83194	6.84
Mediterranean E.	12153	35437	2.92	71395	5.87
Europe	131880	615937	4.67	717879	5.44

Table 3.9   *Continued*

Country or country group	CP/GAI	GAI/CP	Increment
	ratio		%
Finland	1.05	0.96	4.22
Norway	1.86	0.53	3.31
Sweden	1.00	1.00	3.84
Denmark	0.65	1.53	6.85
Germany W.	0.87	1.15	2.35
Germany E.	0.90	1.11	3.51
Poland	1.22	0.82	2.92
Czechoslovakia	0.76	1.32	3.66
Ireland	1.20	0.84	11.00
United Kingdom	1.71	0.58	5.52
Netherlands	0.82	1.22	4.81
Belgium	0.95	1.05	5.20
Luxembourg	0.73	1.37	3.49
France	1.23	0.81	3.96
Austria	0.94	1.07	2.31
Switzerland	1.32	0.75	1.70
Hungary	0.87	1.15	3.99
Romania	0.81	1.23	2.66
Portugal	1.52	0.65	6.83
Spain	0.85	1.18	6.67
Italy	1.75	0.57	2.39
Yugoslavia	1.87	0.54	2.64
Albania			
Bulgaria	1.62	0.62	2.56
Greece	2.84	0.35	2.87
Turkey	2.04	0.49	2.74
Cyprus			
Israel			
Northern	1.11	0.90	3.91
Central	0.94	1.07	2.91
Atlantic	1.60	0.63	6.22
Sub-Atlantic	1.19	0.84	4.03
Alpic	1.02	0.98	2.13
Pannonic	0.82	1.21	2.88
Mediterranean W.	1.03	0.97	6.71
Mediterranean M.	1.82	0.55	2.54
Mediterranean E.	2.01	0.50	2.70
Europe	1.17	0.86	3.34

Table 3.10   *Recorded increments and potential gross annual increment according to Paterson's CVP climatic index and 1950-1990, by countries and their groups*

Country or country group	Potential	1950	1960	1970	1980	1990
				m3/ha		
Finland	3.8	3.0	3.0	3.0	3.3	3.6
Norway	5.3	2.1	2.3	2.0	2.7	2.9
Sweden	4.3	3.0	3.3	3.0	3.3	4.3
Denmark	5.2	6.0	6.2	6.2	9.5	7.9
Germany W.	6.1	4.4	4.9	5.4	5.5	7.0
Germany E.	6.1	5.7	5.3	5.1	6.0	6.8
Poland	5.8	3.1	3.3	4.4	4.0	4.8
Czechoslovakia	6.1	4.6	4.5	4.3	3.8	8.1
Ireland	10.0	3.7	3.4	7.1	7.7	8.4
United Kingdom	8.7	2.2	2.4	3.6	5.7	5.1
Netherlands	6.2	4.1	4.1	4.4	4.2	7.6
Belgium	7.2	3.8	4.1	4.3	7.5	7.6
Luxembourg	6.2	2.1	3.3	4.9	4.1	8.5
France	6.8	3.5	3.6	3.5	4.4	5.5
Austria	6.2	3.4	3.7	6.1	6.1	6.6
Switzerland	7.3	4.9	4.9	5.4	5.7	5.5
Hungary	6.0	4.1	4.2	5.5	6.8	6.9
Romania	4.8	4.7	4.4	4.8	4.9	5.9
Portugal	7.4	2.6	3.0	3.0	4.5	4.9
Spain	3.9	0.8	1.0	3.4	4.6	4.6
Italy	7.1	2.0	2.0	2.7	3.1	4.1
Yugoslavia	6.7	2.4	2.4	3.2	3.3	3.6
Albania						
Bulgaria	5.2	2.3	2.1	2.0	2.3	3.2
Greece	5.3	1.6	1.8	2.2	2.2	1.9
Turkey	6.4	1.7	2.0	2.8	3.0	3.1
Cyprus						
Israel						
Northern	4.25	2.88	3.03	2.87	3.22	3.83
Central	5.97	4.20	4.34	4.81	4.75	6.38
Atlantic	8.90	2.29	2.51	4.18	5.96	5.57
Sub-Atlantic	6.80	3.51	3.64	3.56	4.52	5.69
Alpic	6.47	3.77	4.00	5.93	6.01	6.33
Pannonic	5.04	4.60	4.36	4.94	5.32	6.11
Mediterranean W.	4.83	1.33	1.68	3.27	4.57	4.68
Mediterranean M.	6.84	2.23	2.21	2.98	3.23	3.76
Mediterranean E.	5.87	1.81	1.99	2.47	2.66	2.92
Europe	5.44	2.98	3.09	3.50	3.91	4.67

Table 3.11   *Coniferous percentages of exploitable forests, by countries and their groups, 1990 (for abbreviations, see p. iii)*

Country or country group	Area	GS	NAI	F
	Coniferous percentage			
Finland	92	82	78	78
Norway	77	82	80	97
Sweden	95	85	83	81
Denmark	68	54	73	59
Germany W.	62	67	77	77
Germany E.	75	71	76	78
Poland	79	78	80	76
Czechoslovakia	64	73	70	77
Ireland	90	87	100	98
United Kingdom	72	55	77	82
Netherlands	61	56	61	65
Belgium	49	60	66	67
Luxembourg	78	18	36	49
France	36	38	44	40
Austria	76	83	75	82
Switzerland	69	73	70	74
Hungary	17	15	17	9
Romania	29	40	35	34
Portugal	56	66	61	61
Spain	43	61	69	54
Italy	26	36	31	21
Yugoslavia	13	29	27	27
Albania	17	21	26	12
Bulgaria	36	34	33	32
Greece	41	52	47	28
Turkey	58	72	65	59
Cyprus	99	100	100	94
Israel	57	50	57	80
Northern	91	84	81	81
Central	70	71	76	77
Atlantic	75	59	82	85
Sub-Atlantic	37	39	46	42
Alpic	74	80	74	80
Pannonic	26	36	31	27
Mediterranean W.	47	62	67	57
Mediterranean M.	17	31	28	25
Mediterranean E.	49	58	54	51
Europe	64	64	65	64

Table 3.12   Stand types as percentages of exploitable (EF) and unexploitable forest (UEF) by countries, 1990

Country	Total	High Forest			
		Total	Coniferous		
			Total	EF	UEF
Finland	100	100	92	92	
Norway	100	100	67	67	
Sweden	100	100	95	95	
Denmark	100	100	68	68	
Germany W.	100	98	67	67	
Germany E.	100	100	75	75	0
Poland	100	100	78	77	1
Czechoslovakia	100	98	64	64	
Ireland	100	100	90	90	
United Kingdom	100	100	70	70	
Netherlands	100	91	65	65	
Belgium	100	76	49	49	
Luxembourg	100	85	38		
France	100	50	30	27	3
Austria	100	97	76	76	
Switzerland	100	93	69	50	19
Hungary	100	71	17	16	1
Romania	100	92	29	28	1
Portugal	100	82	56	39	17
Spain	100	87	47		
Italy	100	45	26	15	11
Yugoslavia	100	62	22		
Albania	100	61	23		
Bulgaria	100	73	33	33	
Greece	100	52	39		
Turkey	100	70	58		
Cyprus	100	99	63	63	
Israel	100		57		

Table 3.12   Continued

Country	High Forest			Coppice
	Broadleaved			
	Total	EF	UEF	
Finland	8	8		
Norway	33	33		
Sweden	5	5		
Denmark	32	32		
Germany W.	31	31		2
Germany E.	25	24	1	
Poland	22	22	0	0
Czechoslovakia	34	34		2
Ireland	10	10		
United Kingdom	30	30		
Netherlands	26	26		9
Belgium	27	5	22	24
Luxembourg	47			15
France	20	17	3	50
Austria	21	21		3
Switzerland	24	20	4	7
Hungary	54	51	3	29
Romania	63	61	2	8
Portugal	26		26	12
Spain	40			13
Italy	19	11	8	55
Yugoslavia	40			38
Albania	38			39
Bulgaria	40	40		27
Greece	13			48
Turkey	12			30
Cyprus	36	36		1
Israel	43			

Table 3.13   *Net annual increment (NAI) in relation to fellings (AF),*
percentages by countries and their groups, 1990

Country or country group	100 x NAI/AF
Finland	25
Norway	49
Sweden	58
Denmark	53
Germany W.	60
Germany E.	53
Poland	12
Czechoslovakia	53
Ireland	110
United Kingdom	36
Netherlands	84
Belgium	34
Luxembourg	84
France	37
Austria	27
Switzerland	10
Hungary	36
Romania	98
Portugal	4
Spain	85
Italy	120
Yugoslavia	26
Albania	-39
Bulgaria	87
Greece	-1
Turkey	17
Cyprus	-90
Israel	136
Northern	42
Central	43
Atlantic	48
Sub-Atlantic	38
Alpic	23
Pannonic	81
Mediterranean W.	51
Mediterranean M.	46
Mediterranean E.	26
Europe	43

Austria

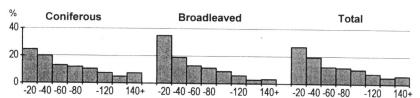

Belgium

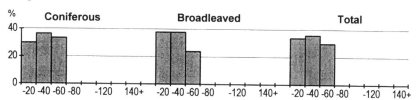

Bulgaria

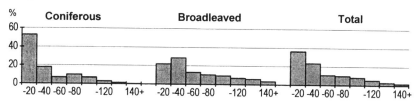

Czechoslovakia

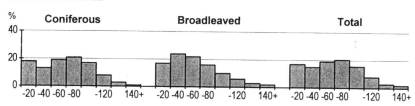

Denmark

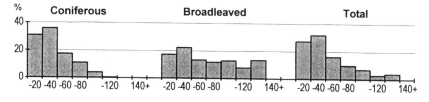

Fig. 3.1   *Age-class distributions (classes 0-20, 21-40, ... and over 140 years)*

Finland

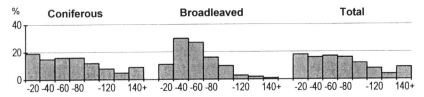

France

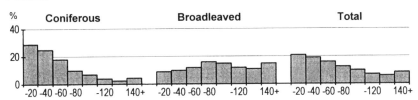

Germany E.

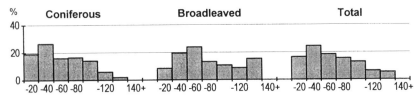

Germany W.

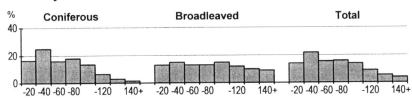

Hungary

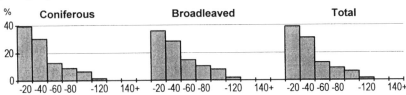

Fig. 3.1   *Continued*

Ireland

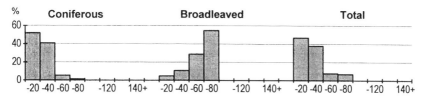

Italy

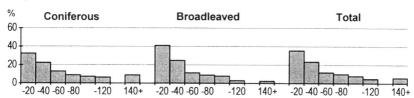

Luxembourg

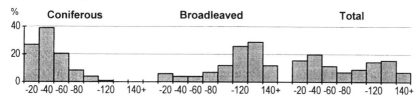

Netherlands

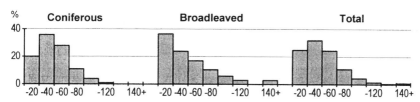

Norway

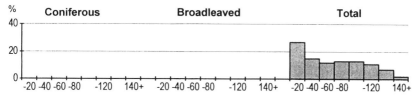

Fig. 3.1   Continued

Poland

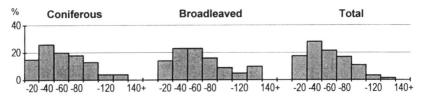

Portugal

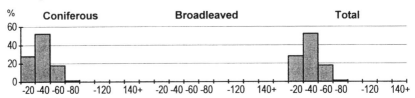

Romania

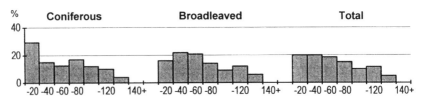

Sweden

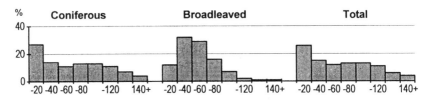

Switzerland

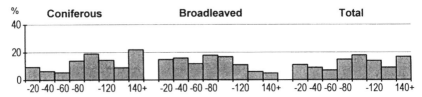

Fig. 3.1   *Continued*

Turkey

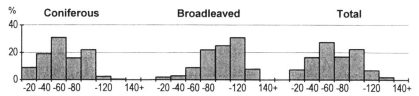

United Kingdom

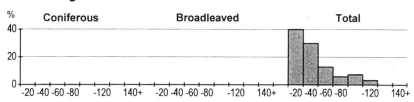

Yugoslavia

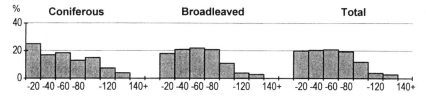

Fig. 3.1   *Continued*

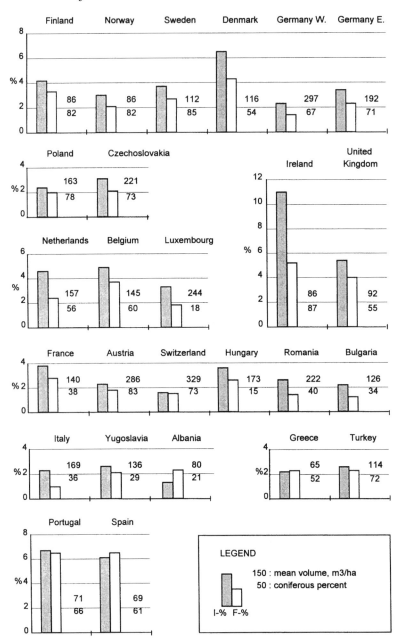

Fig. 3.2   *Annual net increment and annual fellings percents in relation to growing stock (I-% and F-%), growing stock volume per hectare and coniferous percentage (C-%) of growing stock in exploitable forest*

4 ECOLOGICAL AND ECONOMIC BASIS

4.1 Introduction

The natural forest, its tree-species composition and dynamics are functions primarily of climate and secondarily of soil parent materials. Soil, the living upper layer of soil, is created by climate, soil parent materials, flora and fauna. If a fire, storm or insect calamity destroys a tree stand at the climatic climax stage, pioneer plants occupy the site and start a serial succession towards the climatic climax plant community.

Natural forests proper are very rare in present day Europe. The majority of them are in the boreal coniferous zone of European Russia and in small areas of Finland, Sweden and Norway. Only isolated remnants can be found in remote and mostly inaccessible mountain areas in Europe. All other European forests are man-made tree communities maintained by silvicultural and logging measures. Some of them have been degraded by poor logging practices, pasturing and wild fires, often on lands degraded by erosion.

Whatever the current tree-species composition, age and density structure or management regime, there are natural forces inside a forest ecosystem affecting stand dynamics and awaiting the chance to commence succession towards a true climax. Knowledge of the natural forest ecosystems is therefore the basis for analysing the current and future condition of forests and formulating policies and management regimes for developing multi-functional forests which satisfy the needs of man and society (Fig. 4.1, part 1).

The effects of man's activities on forests are dependent on the nature of economy for which the forests have been used. The principal types of economies are presented in Fig. 4.1, part 2. They form the basis for studying and understanding the history of forests and the changing values which forests represent.

Forests have had a commodity function since prehistoric times when man started to gather wood, leaves, fruit, berries, mushrooms, medicinal plants, etc., to satisfy his needs (Fig. 4.1, part 3). The protective function gained importance with advancing industrialisation and an increased density of population. Side by side with the increasing effectiveness of production and transport based on machines and fossil fuels and with a rising living standard, social and cultural functions have gained more importance. Today they predominate in many post-industrial welfare societies.

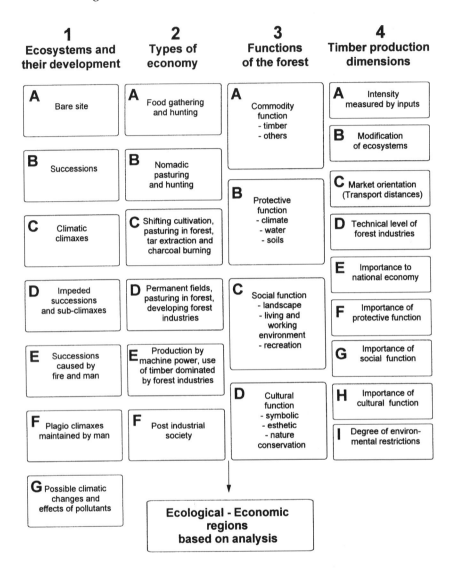

Fig. 4.1   Scheme of analysis of the nature, role and development of forest

The scale of wood production (Fig. 4.1, part 4) has been the physical basis for assessing forest policies. If forestry and forest industries have an important role in the national economy and foreign trade, the commodity function will predominate policy, even in post-industrial societies. On the other hand, if forest resources per capita are small and society is wealthy enough to satisfy the need for forest products by imports, protective, social and cultural functions will be more important than the wood production function.

The analytical scheme presented in Fig. 4.1 is used in this study in order to examine the natural and economic conditions under which forestry is practised. On the basis of the scheme, countries can be placed into groups which belong to the same forest zone or which represent more or less equal stages of economic development. From the point of view of the stability of the current forests, the tree-species composition of the climatic climax stands is the crucial frame of reference. From the point of view of policy implications, the dominating functions of forests and dimensions of wood production are of equal importance.

4.2 *Ecological forest zones*

Climate is the primary factor and soil parent materials the next important factor affecting the natural tree-species composition and functioning of forest ecosystems. A third factor, the water regime, acts as a controller.

The monthly and yearly temperature, the length of the growing season, the amount of solar radiation reaching the ground, the annual amount of precipitation and its seasonal distribution, and especially the ratio of precipitation and evaporation during the growing season are the most important climatic factors. An example of the applicability of climatic factors in estimating the potential yield of stem wood is the CVP index developed by Paterson. The independent variables of the index are as follows: mean temperature of the warmest month, range between the mean temperature of the warmest and coldest month, mean annual precipitation, the growing season in months and the evapotranspiration reducer (see section 2.8 and fig. 4.2)

Europe is located between the latitudes of 70°N and 35°N, i.e. from the Arctic to the Mediterranean. The cold climate and short growing season in the North and evaporation greater than precipitation during the growing season restrict the growth of trees and determine the tree-species composition.

The mean annual temperature of July ranges from +5 to 30°C, that of January from minus 15 to +10°C, and the annual precipitation from 300 to 2000 mm. Outside the closed forest there is forest tundra and treeless tundra in the North, forest steppe and grass steppe in the South-East, and subtropical scrub land and spots of desert in the South.

Europe can be therefore divided into three principal forest zones (Fig. 1.1):

Ia - Boreal coniferous forest in the cold Northern latitudes.

II - Broadleaved deciduous forest in temperate zones.

III - Evergreen mixed forest in the Mediterranean zone.

Between the boreal coniferous and broadleaved deciduous forest zones there is a latitudinal ecotonal, I-IIa, and on the slopes of mountains an elevational ecotonal mixed forest. Above the latter, there is a Sub-Alpine coniferous forest, Ib, below Alpine tundra and barren mountain tops.

4.2.1 Boreal coniferous forest (Ia)

The temperature and length of the growing season are the principal climatic variables which determine the growth of trees in the boreal coniferous zone. The July 13 °C isotherm is located approximately on the Northern boundary of the closed forest. Changes of climate and wild fires have over time shifted the forest boundary to the north and the south. The southern boundary facing the ecotonal mixed forest approximately coincides with the July 18 °C isotherm.

Mountains, oceans and seas, and maritime and continental climatic influences cause deviations of the zonal boundaries from the isotherms. Climates warmer than average push the southern boundary northwards, and climates colder and drier than average move the northern boundary southwards. Maritime influences predominate in the western regions adjacent to the Atlantic Ocean, while continental influences increase towards the east.

The temperature of the warmest month ranges from +10 to 15 °C and that of the coldest month from +2 to minus 3 °C in the maritime part of the zone. The climatic extremes are greater in continental than in maritime regions. The monthly temperature varies considerably in the northeast. Desiccating winds together with temperatures of minus 20 to minus 40 °C destroy needles and can be lethal to trees near the northern forest boundary.

Warming in the spring can be rapid. Spring frost cause damages to spruce, as well as to trees transferred from more southernly provenances and exotics. The date when photosynthesis begins varies greatly. Summer weather is changeable and the length of the growing season, based on a daily temperature over +10 °C, varies from 30 to 120 days. The light regime differs greatly from that of lower latitudes. Days are long during the growing season.

The annual precipitation varies from 400 mm in the north to 800 mm in the south, and it can be as much as 2000 mm on the Atlantic coast. Much of the precipitation falls in the form of snow. The winter is mild in the west, with the cold increasing towards the east. The snow cover lasts from 4 to 7 months.

A great deal of precipitation falls in the growing season. In the cool and humid climate evaporation is less than precipitation except during the warmest times in the southern parts of the zone. Shortages of water do not restrict the growth of trees except in periods that are drier than average. However, the climate promotes paludification and the development of peat-forming mires. In some parts of the zone wetlands cover 60 to 70% of the land area.

Glacial and alluvial sands, gravels and moraines are the commonest soil parent materials in the boreal zone. In the cool and humid climate, soil profiles are often podzolic. Podzolisation is fastest in sandy glacial outwashes. In more fertile soils the process is slower and profiles are nearer to brown soils than podzols proper. They are referred to as brown podzols. Soils and their parent materials are less fertile than those in ecotonal mixed and deciduous broadleaved zones. The ability of soils to buffer acidification is considered to be relatively weak, although the buffering mechanisms of the naturally acidic boreal soils are not yet fully known.

Boreal trees are adapted to a fire ecology and for the invasion of treeless sites. Under natural conditions, wild fires ignited by lightning are repeated, on the average, approximately once in 50 years on dry sandy and gravel sites. The time between fires on moist sites is 100 to 150 years, and over 200 years on the richest and moistest sites.

Birches, aspen and alder and also pine on dry sands and gravels are pioneer trees which invade burnt sites (Table 4.1). Spruce is the only climax tree in Northern Europe. Its seedlings develop under the pioneer trees and occupy the site step by step. Fire maintains the rotation of pioneer and climax stands, the fertility of sites and the biodiversity of forests.

Table 4.1   Native tree species of European forest zones, exotics and hybrids

I a Boreal Coniferous		I-II Mixed ecotonal a) latitudinal b) elevational	
Pinus silvestris	Scots pine	*Pinus silvestris*	
Picea abies	Norway spruce	*Picea abies*	
Betula pendula	white birch	*Betula spp.*	
Betula pubescens	downy birch	*Quercus robur*	pendulate oak
Populus tremula	aspen	*Fagus silvatica*	beech
Alnus incana	grey alder		
I b Sub-Alpine coniferous --->			
Exotics --->			*Pinus contorta*
Pinus contorta	lodgepole pine		*Picea sitchensis*
			Pseudotsuga taxifolia
			Robinia pseudiacacia
Hybrids --->			
			Poplars,

Table 4.1   *Continued*

II Broadleaved deciduous		Evergreen mixed	
Atlantic and Sub-Atlantic Europe		*Quercus ilex*	holm oak
Quercus robur		*Quercus suber*	cork oak
Quercus petraea	sessile oak	*Quercus coccifera*	Kermes oak
Fagus silvatica		*Quercus pubescens*	
Fraxinus excelsior	ash	*Castanea sativa*	
Ulmus spp.	elm	*Olea Europea*	Olive tree
Tilia cordata	lime	*Alnus cordata*	Italian alder
Pinus silvestris		*Cypressus sempervirens*	Italian cypress
Central Europe		*Pinus pinaster*	maritime pine
Quercus robur		*Pinus pinea*	stone pine
Fagus silvatica			
Carpinus betulus	hornbeam		
Castanea sativa	sweet chestnut		
Southern Europe			
Quercus Lusitanica			
Quercus cerris	Turkey oak		
Quercus pubescens	downy oak		
Acer platanoides	European maple		
Fagus silvativa			
Castanea sativa			
Fraxinus cornus	manna ash		
Pinus silvestris		*Pinus silvestris*	
Picea abies		*Pinus nigra*	black pine
Abies alba	silver fir	*Pinus peuce*	Makedonian pine
Larix decidua	European Larch	*Pinus halepensis*	Aleppo pine
Pinus cempra	cembran pine	*Abies alba*	
Pinus mugo	mountain pine	*Abies pinsapo*	Spanish fir
		Abies cephalonica	Greek fir
		Abies cilicica	Cilician fir
		Cedrus libani	Lebanon cedar
		Pinus radiata	radiata pine
		Eucalyptus spp.	
Sitka spruce			
Douglas fir			
false acacia			
Larches		Poplars	

For all practical purposes, erosion is unknown as an agent impoverishing boreal soils. Even after the fiercest fire, or on clear-cut areas, the fast growing ground vegetation retains the bulk of the released nutrients in the nutrient cycle, making them available to trees in later stages of the succession. Humus and plant roots also help to prevent erosion.

The boreal forest is one of the world's most durable plant communities. It can only be destroyed by land use change or by exceedingly high emissions from mining and metallurgical industries.

The qualities of boreal trees also illustrate the natural development of tree-species compositions in ecotonal mixed and Sub-Alpine zones.

Conifers are well adapted to cold climates. They reduce transpiration of water during the winter when water cannot be obtained from the frozen ground. Occasionally, however, they suffer needle losses in extremely cold weather with desiccating winds or in exceptional weather in early spring when mild winds initiate transpiration from needles but the trees cannot obtain water from the still frozen ground.

The photosynthesis of conifers begins in the spring at a lower temperature than that of broadleaved trees. Additionally, conifers have needles which permit photosynthesis as soon as the temperature allows it while broadleaved trees have first to grow their leaves, losing therefore part of the growing season. Broadleaved deciduous trees have to use a part of their net primary production to grow their leaves every spring while conifers renew their needles once every 4 to 7 years.

Conifers grow to a greater height, produce a greater growing stock volume per hectare and reach higher ages than broadleaved trees. Because of these qualities they can outgrow broadleaved trees. The shade tolerance of spruce is also an advantage in competition.

The competing power of broadleaved trees is based on their plentiful crop of seeds transported by the wind, their capability to regenerate by vegetative sprouting and their fast initial growth. Broadleaved trees become more competitive with regard to conifers as the temperature increases.

The average range of the potential annual yield of stem wood, over bark, is approximately 2 to 5 m^3/ha. It is 1 to 2 m^3/ha on the most barren sites and 6 to 8 m^3/ha on the best sites in those parts of the boreal zone where closed forest is possible (Fig. 4.2). On the sites of equal quality, the volume yield of conifers is greater than that of broadleaved trees.

Fig. 4.2   *Potential annual yield of stem wood, m³/ha, over bark, on*
               the basis of Paterson's climatic index by European countries

4.2.2 *Broadleaved deciduous forest (II)*

Broadleaved deciduous (summergreen) forest is the natural vegetation
on most Europe between the latitudes 40° N to 60° N and longitudes 10°
W to 45° E. Radiation, light and temperature regimes changing from the
north to the south, maritime influences in the west and continental
influences in the east determine the growing conditions.

The mean annual temperature is approximately +10 °C. The July mean
air temperature varies from 15 °C in the maritime northwest to more
than 20 °C in the continental southeast where the closed forest borders
forest steppe. Summers are moderately warm and the growing season is
interrupted by mild winters lasting 2 to 4 months. The January mean
temperature varies from +5 °C in the west to minus 5 °C in the east. The

winter cold restricts the growth of tree species transferred from warmer vegetation zones.

Respiration and consumption of assimilated products continue at a low level during most of the mild winter. Because of this, the yield of stem wood is not much greater than yields at higher latitudes despite the temperature differences.

Mean annual precipitation varies from 1000 mm on the Atlantic coast to 600 mm in the east. The extremes are 2000 and 200 mm. Much of the precipitation falls in the summer months.

From the point of view of tree growth, the difference between annual precipitation and potential evaporation is more important than the amount of precipitation as such. Evaporation varies from 750 mm on the western coast of Atlantic Europe to 250 mm in Central Europe and to minus 250 mm in the southeast. At times of high temperature and low humidity the shortage of water decreases tree growth and causes defoliation.

Winds can be strong. Storms, especially in the Atlantic part of the zone, and even in central regions, cause damage unknown in the boreal zone.

Soil parent materials are varied and richer than in the Boreal zone. Brown and grey soil profiles predominate. Layers from many geological periods often form micro-mosaics of sites. In the northeastern and northern parts of the Central European plains, the soil profiles are podzolic under oak-birch forests. The annual fall of leaves and other litter is considerable. The ability of soils to buffer acidification is better than in the boreal zone.

In the Atlantic part of the zone, the range of temperature is small, climate is moist-temperate, microclimates inside the forest are shady, and beech is often a climax tree. The natural tree-species composition is generally a mixture of deciduous broadleaved trees (Table 4.1).

In the eastern part of the zone, the microclimates are less shady, even sunny on some sites. Consequently oak is a characteristic tree, often accompanied by pine. Under natural conditions, broadleaved trees predominated in the zone. Pine may have been a common tree on sandy outwashes, possibly because of repeated wild fires.

The change of environmental factors from the boreal zone to the broadleaved deciduous zone increases the competitive ability of broadleaved trees over conifers. Deciduous and evergreen strategies are well illustrated by comparing beech with spruce.

Beech is able to regenerate and thrive better in the shade of trees other than spruce. It is less sensitive to storm and snow damage than spruce because of its leafless crown in the winter (when icy snow accumulates on tree crowns). Spruce is often windthrown because of its horizontal root system. Beech lives longer than spruce which often suffers from fungal diseases.

On the other hand, the net primary production and the volume yield of wood are both greater in spruce and other coniferous stands than in broadleaved stands on the same site quality. Conifers have a greater photosynthesising biomass than broadleaved trees, while other factors causing yield differences are discussed in the previous section (4.2.1) concerning the boreal zone.

The potential annual yield of stem wood, over bark, in the broadleaved deciduous forest, is at least 5 to 6 m³/ha. On the best sites it is 10 to 15 m³/ha. Yield is greatest in the maritime part of the zone where precipitation is abundant and occurs throughout the year, being greatest on mountain slopes where there is a better water regime than in neighbouring lowlands. Yields are smallest in the southeast where potential evaporation exceeds precipitation (Fig. 4.2).

4.2.3   Mixed evergreen forest (III)

Mixed evergreen forests are located around the Mediterranean Sea and the Black Sea. Typically, in olive climate, summers are warm or hot and winters mild. The mean temperature of the warmest month ranges from +10 to 35 °C and that of the coldest month from +2 to 18 °C. Winter frosts may occur but they do not prevent the growth of exotics such as eucalyptus.

The mean annual precipitation varies generally from 500 to 800 mm, but it is 800 to 1000 mm on the western coast of the Adriatic sea and 200 to 300 mm in parts of the inland Spain and Turkey. Precipitation falls mostly in winter, occasionally as snow in lowlands and regularly in the mountains and the northern parts of the zone.

Annual precipitation minus potential evaporation in the Mediterranean differs greatly from that in the broadleaved deciduous forest. It is the climatic factor most effectively restricting the growth of trees. The difference is smallest or its negative value is greatest in the southern inland regions of Spain, in the southernmost part of Italy, in Greece and in inland Turkey. Treeless steppe and desert are the natural vegetation types in the driest areas.

There are areas in the Mediterranean countries where the climate differs considerably from that of the Mediterranean region proper. On the mountain slopes of the Iberian peninsula facing the Atlantic Ocean the climate is sub-tropically warm and moist. The water regimes of the Sub-Alpine slopes in northern Italy and in the Dinaric Alps of Yugoslavia support yields of wood greater than the average for the zone as a whole.

Water availability is the main factor limiting tree growth. The zone is divided into sub-zones on the basis of the number of biologically dry days during the dry season. The numbers of rainy days and days with mist or dew during the dry periods are also important variables.

Only fragments remain of the rich and continuous evergreen forests which covered the Mediterranean as true climatic climaxes at the beginning of recorded history. Consequently, it has been difficult, if not impossible, to determine the tree-species composition of the original natural forests.

Rocky sites supporting either grasses and scrub or devoid of vegetation predominate the current landscape outside cultivated fields, pastures, gardens and the scattered forests proper. There is evidence that in addition to forests, a small part of the current maquis originally formed climatic climax plant communities.

Evergreen oaks are characteristic natural trees in the lowlands (Table 4.1).

Holm oak is common except in the eastern Mediterranean while cork oak is common in the west close to the sea. Stone pine grows on scattered sites and in small groups throughout the area up to the elevation of about 450 m; similarly Aleppo pine in the east. Italian cypress, sweet chestnut and olive trees are also characteristic of the region.

In the original forests, soil types, ground vegetation and tree regeneration strategies had developed for the shady and moist conditions under the canopy of tree crowns. Direct sunlight reached the ground only in the openings left by dead trees. New seedlings filled the openings. Humus and soil were cool and moist even in the hottest weather.

Disturbances revealed the ground surface to direct sunlight and desiccating wind. If the openings were large enough, microclimates, soils and water regimes were destroyed and soils with nutrients were swept away by erosion. The current sites with red soils, rocky slopes and ravines, bare rocks and scrub maquis seriously restrict tree regeneration.

The potential yield of stem wood, over bark, is determined by the water regimes, with good averages being from 4 to 6 m^3/ha. Yields barely reach 2 to 4 m^3/ha in the closed forests of the driest climate. On the other hand, the greatest yields can be about 15 m³/ha on irrigated sites and on the Atlantic coast (Fig. 4.2).

4.2.4 Ecotonal mixed forest (I-II a,b)

Ecotonal mixed forests form a transitional sub-zone between coniferous boreal and sub-Alpine forests and broadleaved deciduous forests. The latitudinal mixed forest is located south of the boreal zone and the elevational mixed forest downslope from the Sub-Alpine zone. In the transition belt, conifers and broadleaved trees are equally competitive. A detailed reconstruction of the original distribution of the mixed forests is impossible because man has greatly reduced its extent and altered its original tree-species composition.

Latitudinal mixed forests are found in Scotland, southern Norway and Sweden and in the eastern parts of Central Europe. In the lowlands towards west and south, the dominance of broadleaved trees such as beech increases under natural conditions (Table 4.1). Pedunculate oak was a common tree in the east. Scots pine grew on outwash sands in the northern and eastern parts of Central Europe, while Norway spruce was most competitive towards north.

Elevational mixed forest occurred on the slopes of both the Alps and the Carpathian, Balkan and Catalonian mountains. Norway spruce was probably more competitive in the elevational ecotone than in the lowlands.

The light regime, length of growing season and occurrence of spring frost in the southern and northern parts of the ecotonal zone differ considerably, which should be taken into account when transferring trees from one locality to another.

4.2.5 Sub-Alpine coniferous forest (Ib)

Sub-Alpine forests are coniferous forests which grow in climatic conditions, which except for the light regime and the length of day, closely resembles that of the Boreal coniferous forest zone. Remnants of the original Sub-Alpine forests can be found in certain inaccessible locations on the mountains of the southern half of Europe.

In addition to Scots pine and Norway spruce, silver fir, European larch and Cembran pine are the natural conifers of the elevational ecotone within the broadleaved deciduous zone. The Sub-Alpine forests on the mountains within or near the evergreen mixed forest zone have their own suite of conifers (Table 4.1).

The existing remnants of Sub-Alpine forests are mostly classified as national parks, natural reserves or other types of unexploitable forest. They form a rich genetic reserve and many of them are planted for landscape and aesthetic purposes.

4.3   Human influence on the development of European forests

Under natural conditions before the influence of man forest covered 70 to 80% of the land area in Europe. The coverage was near 100% on the central European plains and hills. Only the highest mountains, some wetlands, semi-deserts and barren moorlands were not under closed forest. The forest coverage is now 27% and only 5 to 10% in the most densely populated areas.

During the last 2000 years, 4000 years in the Mediterranean, man has drastically exploited forests, cleared them for fields, pastures and built-up areas, or reduced forest to scrub land. Most of the current forest ecosystems are quite different from those of the natural forests. They are at best plagio-climaxes maintained by silvicultural and logging measures. Their tree-species composition, density, age structure and biodiversity are artificial. The last natural forests untouched by logging can only be found in Northern Europe and as scattered remnants on the highest mountains in other parts of Europe.

4.3.1   Northern Europe

Compared with other regions of Europe, Northern Europe was settled relatively late, mainly during the last 1000 years. Podzolic soils made poor fields during the time of manual labour. The rigorous conditions prevented the growth of population into numbers whose demand for wood would have been sufficient to destroy the forests. The boreal trees' regenerative vigour and ability to reinvade abandoned and neglected fields and pastures maintained the forest coverage.

Advancing settlements were only possible with the support of wood resources. Wood was used for buildings, heating, cooking, tools, and for transportation, especially boats and ships. Shifting cultivation and pasturing in the forest permitted settlement to penetrate deeply into the wilderness.

Mining and shipbuilding became great consumers of wood during the 17th century. Charcoal burning and tar extraction offered work and income until the end of the 19th century. The increasing population exploited the forests wastefully. According to records, wood resources reached their lowest point in the middle of the 19th century. Regulations were then put into force to protect the forests and to ensure their regeneration after final cutting, despite which destructive forms of selection cutting continued until the first half of the 20th century.

Together with the increasing demand for industrial wood, the practice of shifting cultivation and pasturing in forests diminished, other sources of energy substituted fuelwood and the wasteful consumption of wood by the rural population decreased. The standwise cultivation of trees, thinnings and cuttings proper became economically profitable with the demand created by the pulp industry, with a consequent development of stumpage prices for all wood assortments.

The hand saw and axe were replaced by the chain saw, and the horse by the forest tractor and truck, and finally, by the harvester that fells and debranches trees, and prepares and measures wood assortments. The forest tractor made the complete mechanisation of logging and thinning possible.

Forest improvement measures and intensified silviculture were employed when the increasing demand for industrial wood threatened to lead to overcutting in the 1960s and 1970s. Silvicultural inputs added to the increment of growing stock so much that for the first time in last 200 years in Northern Europe, fellings are currently much smaller than the sustainable cut possibilities.

4.3.2 Central Europe

In Central Europe forest degradation and the clearing of forest for fields and pastures started earlier and, because of the dense population, led to a much greater decrease of forest resources than in Northern Europe. Selection cuttings, high consumption of wood by the farming population, shifting cultivation and intensive pasturing, e.g. by sheep in Atlantic Europe, shipbuilding, the consumption for fuel and the use of wood in the mining and metallurgical industries decreased forests to a minimum in the 18th and 19th centuries.

In shifting cultivation, trees are cut and burned on the site. The land is farmed for 2 to 4 years and then left to reforest. This method made an extensive use of land. It required many times more area than current agriculture to produce an equal crop. The rotation was c. 10 to 20 years. When combined with the production of fuel and utility wood, the rotation was c. 30 to 40 years.

Suitable tree species for shifting cultivation were those broadleaved trees, such as oak, beech, hornbeam and alder, which regenerate well by stump and root sprouts. Coppice is a management regime based on sprout regeneration. The method has been an important producer of fuelwood and timber for fences, orchards, vineyards and household constructions. Coppice with standards, i.e. with a small number of high forest trees, produced both small and large-sized timber.

The leather industry also increased the amount of coppice to produce tannin bark. The demand and price for bark was so high in the 18th and 19th centuries that high forests and even fields were converted to oak coppice. The last short-lived booms for tannin bark were during the two World Wars of the 20th century.

Browsing, a type of pasturing in which swine graze on acorns and beech nuts, was also practised in Central Europe, although it has been and still is more important in the Iberian peninsula.

Decreasing and degrading forests led to a shortage of wood and the loss of the protective benefits of forest coverage. In the pre-industrial era wood was also harvested in the mountains up to the boundary of the closed forest. Erosion, landslides and avalanches caused losses of soil, floods and damage to field, buildings and roads.

The idea of a regulated and sustained-yield forestry was first presented in the 16th century in Central Europe and was further developed on the basis of research and experience in the 18th and 19th centuries.

Sustained-yield management became legally binding in the German states during the period 1833-1900. The target was a sustained yield of wood, and its means were regeneration after final harvest, the tending of seedling stands, silvicultural thinning to harvest wood otherwise lost by mortality and growing the improved, fully dense stands to maturity and final harvest. Another target was to afforest lands growing scrub vegetation and to convert coppice into high forest.

Managed forestry was first practised on royal hunting grounds protected from general utilization. The combined interest of hunting and sustainable forestry explains the high status of hunting as a non-wood function of forests in Central Europe at the present time.

States, municipalities and owners of large private estates also initiated sustainable management practices. On the other hand, the material needs of the growing agrarian population maintained over-exploitation in those forests in which they had the right to harvest wood and fodder and pasture their animals.

Technological progress and industrialization, as well as the intensification and rationalization of agriculture and the substitution of fuel and household wood by other sources of energy and materials, permitted the afforestation of marginal fields and pastures as well as the conversion of scrub vegetation and coppice into high forest. The increased demand for industrial wood made silviculture and forest improvement profitable. The greater value of coniferous wood as a raw material for industry compared with broadleaved wood led to the favouring of conifers in afforestation and improvement planting.

There have been, and still are, economic and social obstacles to converting coppice into high forest. The harvesting and regeneration costs of coppicing are higher than the market price for small-sized broadleaved wood. Most of the remaining coppice consists of small lots owned by farmers and other private people. Policy measures and subsidies are needed to convert them into high forests.

Decreasing consumption of fuel and household wood has led to a considerable accumulation of the growing stock to amounts which have been under-estimated with respect to both stock and increment. In spite of considerable imports of forest products by European countries, their industrial capacity to use their wood resources fully has been and still is too small. The low profitability of logging and the increasing importance of environmental functions of forests have led to longer and longer rotations and increasing stand density. Consequently, current removals are much smaller than the potential removals permitted by sustainable production.

4.3.3 The Mediterranean

At the beginning of recorded history, evergreen mixed forests largely covered the Mediterranean landscape. Most of these ancient forests have vanished. The current landscape serves as an example of the consequences of man's unconsidered activities. Nevertheless, warming of the macroclimate together with the decrease of precipitation seems to be the major reason for the disappearance of the ancient forests. Man's activities have accelerated the natural processes by exploiting forests with methods which made the natural regeneration of trees difficult or impossible.

The early Mediterranean cultures used wood for buildings, heating, tools, implements and, above all, shipbuilding. They were also seafaring nations and the Greeks, Phoenicians, Romans, Turks, Venetians, Spaniards and Portuguese traded throughout the Mediterranean and later as far as the New World using wooden ships.

Wood was exploited by methods which revealed the ground to direct sunshine and desiccating winds thereby destroying the natural soil and water regimes. The ecological basis for forest growth was destroyed. The ancient attitudes were in fact hostile towards forests. Fields, pastures and gardens were valued as cultural landscapes in contrast to the disorderly wilderness. This attitude is still common and constitutes a major threat to forests and reforestation projects.

Shifting cultivation and wild fires have opened the closed forests. Pasturing, especially of goats, has prevented regeneration. Wild fire still takes a heavy toll in forests and afforestation plantations.

Traditionally, forests have been multi-functional. Growing populations have harvested wood and gathered other forest products to meet their needs. They also pastured the forests more or less freely. These practices still continue in large areas in private and public forests.

While pasturing has destroyed forests, in some cases it has also maintained them. Since Roman times, swine browsing on mast, i.e. by acorns and beech nuts, has been important. Mast is produced by trees growing in open stands where the distance between trees is about 15 m. Trees are pruned to increase the mast crop.

Mast and cork have been a profitable combination. Cork is one of the major export products in Portugal and it is exported to a minor extent also in Spain and Italy.

The production of charcoal consumes a great deal of wood, although its transportation is much cheaper because it is lighter than wood. The use of charcoal for fuel has a long history, and the tradition still continues in some regions where it can be a serious drain on wood resources.

Hunting has traditionally had a high status in Spain and other parts of the Mediterranean. Measures to maintain the old royal hunting grounds and to protect them also protected the forests.

The traditional multiple-use of forests hinders efforts to grow closed forest for production of industrial wood. In closed production forests, pasturing and gathering wood for household purposes are excluded. Similarly, fast growing commercial stands, especially those of eucalyptus, may consume scanty water resources, decreasing the availability of water for other uses. Changes in the cultural landscape are also unpopular.

Portugal is an example in the Mediterranean region where forest as a producer of industrial wood has a high status side by side with multi-product forests. On the other hand, in Spain pine plantations which cover approximately 3.5 million ha are not wood production forests proper. They have been established in order to prevent fluvial erosion.

The traditional use of forests and the attitudes of the fairly large farming population remain obstacles to projects aimed at improving and increasing forest resources, thus the degradation of forests continues in large areas of the Mediterranean East.

4.4   Forest management regimes

4.4.1   Basics of forest management regimes

A forest management regime consists of a prescribed tree-species composition and measures for growing trees and harvesting them for timber. The measures can be divided into four groups: stand establishment, tending the seedling stands, thinnings producing usable wood, and final cuttings at the prescribed rotation age of the trees.

Tree-species composition on different site qualities determines the productivity of stands for a long time. In cases of seeding by mature trees, the tree-species composition of the earlier generation determines the species of the new stands. The composition, however, can be regulated by tending measures which favour a desired crop composition. The freedom to choose species is greater when regeneration or afforestation is artificial seeding or planting.

Various thinning regimes can be applied to the growing stock. The intensity is the average annual volume per hectare removed over a period of one or more thinning cycles. A cycle is the number of years between successive thinnings. The weight is the volume of trees removed in a particular thinning, while type of thinning describes its qualitative nature; while it may be a low or high thinning.

In the final cutting, all mature trees can be removed by one logging operation, by clear-cutting, or by two or more operations. Gradual cutting is employed when the shelterwood or seedtree methods are used for obtaining natural regeneration. The rotation is the period of time between successive regenerations. In the case of gradual cutting the regeneration begins under the maturing or mature trees. Rotation is then somewhat shorter than maturity age.

In coppice management, mature trees are clear-cut and a new stand regenerated by vegetative sprouts. In coppice with standards, the coppice rotation is shorter than the rotation of standards, which are regenerated by seeds. When the stand is regenerated by seeds, it is called high forest management.

Selection cutting management is used in limited areas such as experimental forests, as well as on the mountains and in forests managed for landscape, protection and recreation. Regeneration, the tending of seedlings, thinnings and the cutting of mature trees is carried out on the same area, so the stand is composed of trees of different ages and sizes.

The objective in defining and prescribing management regimes has traditionally been the highest possible net income per hectare per year for growing, harvesting and selling wood. Another criterion of a profitable management regime is to maintain a rate of interest higher than the opportunity cost on the capital invested in forest or represented by the growing stock. Although products other than wood, i.e. material and non-material benefits, have gained in importance, the economic criteria of wood production have been the principal bases for prescribing management regimes. Physical and technical standards such as dimensions of crop trees are often used as substitutes for the more complex economic standards.

Net annual income, or forest rent, as a variable of profitability can be applied to a single stand or a forest composed of a number of stands. It is the surplus of the average annual gross income over annual expenses. The interest on the capital invested into silvicultural inputs or tied up in the growing stock is not considered.

Net discounted revenue, land rent, and internal rate of return measure the standwise economic success of silvicultural inputs, such as the choice of tree species and different ways of establishing stands, and different regimes for thinning and harvesting stands. The interest as the cost of capital measures the benefits lost between the times of investment and of revenue. It also measures the opportunity cost of the capital invested into forestry with respect to alternative investments available to forest owners.

The marginal rate of return is an economic criterion applied to the management of the existing growing stock as accumulated capital convertible into money under the restrictions of the opportunity cost of the capital. It is estimated by the marginal analysis of the input/output relationship where the increase of net income is compared to the required increase of growing stock and expressed as a percentage.

The following relationships between the economic criteria are worth noting. The marginal rate of return is zero in a fully regulated growing stock managed by the rotation of the maximum net annual income. The rotation of the growing stock of a fully regulated forest managed at a prescribed marginal rate of return percentage approximately equals the rotation of the maximum net discounted revenue which is calculated by the same percentage which is the prescribed marginal rate of return in a standwise analysis.

The economic criteria of management regimes have been developed for growing stands composed of trees which are more or less of the same age. Net annual income and marginal rate of return can also be applied to selection cutting management.

Because of the long time required to grow seedlings into a mature crop, net discounted revenue and internal rate of return can be used beneficially to estimate the profitability of alternative regimes and inputs into site improvements. When the accepted regimes are in use, the time of investments such as artificial regeneration, afforestation and site improvements are several decades in the past, the investments are often considered as sunken costs with little relevance for decisions on the present treatment of growing stock. The situation in the case of fast growing trees such as eucalyptus and poplar is different, and discounted revenue and internal rate of return can both be used as criteria of profitability.

With a growing stock of long rotation trees, the relevant criteria of economic profitability are net annual income and marginal rate of return. Net annual return has a disadvantage caused by the relatively small operation costs compared with the gross income. Losses caused by seemingly small increases of operation costs can accumulate unnoticed into amounts which significantly decrease profitability. The marginal rate of return is an efficient, though little used, criterion of economic success in cases of large growing stock volumes. It can be used for estimating those stages of growing stock at which the marginal rate of return is zero or negative causing inevitable economic losses in forests where the wood-production function predominates.

The economic criteria described above will be modified for forests where functions other than that of wood production predominate. In the case of non-material benefit functions, the success of management has to be based on value judgments.

The following regional review discusses only those features of management regimes which are relevant to describing the current conditions of European forests. The review concentrates on those characteristics which form a basis for decisions concerning future treatment.

4.4.2 Northern Europe

The growing of even-aged stands is the only profitable way to produce wood in the fire ecology of the boreal coniferous zone and with the native tree species of Scots pine, Norway spruce and birches. Grey alder and aspen are of little economic value. Their proportion is 2 to 3% of the growing stock. Oak and beech can be grown in the southernmost parts of Sweden and Norway. Their value as an environmental asset is greater than as a producer of wood. Lodgepole pine is the only exotic in production forests.

Most of the current stands have been established by natural seeding. Artificial planting and seeding have increased during recent decades, and today account for 70 to 80% of the area regenerated.

Pine is grown on the sites of poor and medium quality, spruce and birch on the sites of medium and better quality. In order to guarantee the profitability of forestry, there is usually one species which predominates with one or two species as mixtures. Nowadays a mix of 10 to 20% secondary species is recommended, especially of birch. The aim of the mix is to maintain the site and stand productivity, the health of the stand and its environmental value.

A prescribed rotation is 80 to 120 years for pine and 70 to 90 years for spruce in the southern part of the region and of 110 to 180 and 90 to 140 years, respectively, in the northern parts.

In the existing virgin forests, many stands may be much older than the prescribed rotation age. Because felling volumes are less than the net increment, the density and age of stands in large areas exceed those prescribed. Actual rotations are often 10 to 20 years longer than the prescribed ones. Mature and over-mature stands account for 20 to 30% of the forest area. Where there are over-mature and degrading stands, the marginal rate of return is therefore smaller than for a fully regulated growing stock.

4.4.3 Central and Alpic Europe

The management of even-aged stands predominates in the region. Selection cutting management is practised in small areas of production forests in the mountains where spruce, fir and beech are the natural, shade-tolerant trees. Coppice and coppice with standards are grown in relatively small areas. Tree species are mostly of local transfers, and there are a few exotics such as the Douglas fir.

In the 18th and 19th centuries, much of the low-quality broadleaved forest was replaced by large coniferous reforestation projects. The bulk of Central and Alpic European forests on lowlands are now composed of coniferous trees in climatic conditions where the oak and beech are the natural climax trees.

Planting remains the most common method for establishing new stands. The annual plantations amounted to 1.4% of the large closed forest areas during the period 1950-1974 when the areas clear-cut during and after the World War II were regenerated. Conifers accounted for 70 to 80% of the total area planted.

The number of seedlings planted has traditionally been 5 to 7 thousand per hectare in the case of spruce and 10 to 20 thousand in the case of pine. The high number of seedlings has been justified by the target to grow high quality wood by self-pruning in the young stands. Small-sized poles and rods thinned from such stands have earlier been valuable assortments for fences, gardens and vineyards. The tradition dates back to the times when labour was relatively much cheaper than today. There is no economic justification for the continuation of this practice. The costs of the large number of seedlings per hectare are one of the major factors decreasing the profitability of forestry.

Prescribed thinning cycles have been 5 to 10 years and the weight of thinning relatively small. The rising logging costs and also an insufficient demand for small-sized timber have postponed thinning, resulting in over-dense stands.

In the 19th century and at the beginning of the 20th, rotations were based on the net discounted revenue. This criterion led to relatively short rotations and small growing stock volumes per hectare. The high afforestation investments proved to be sunken costs. Net annual income and wood reserves as such were considered to be the criteria of success. Today, rotations are lengthening.

In the 1920s and 1930s a prescribed rotation for most conifers was about 80 years and for fir 90 years. In the 1980s, it was 100 to 120 years for pine and larch, 80 to 100 years for spruce and fir, 140 to 180 years for oak and 120 to 140 years for beech in the western part of Germany. The actual rotation for conifers is 110 to 120 years, 120 to 130 years in the state forests. In Poland, the former Czechoslovakia and the eastern part of Germany the rotations are somewhat shorter. In Switzerland, 22% of coniferous forests are older than 140 years.

There are two principal reasons for the increasing length of rotation. In welfare countries there has been no urgent need to rationalise and mechanise forest work. Increased demand for forest products has been met by increased imports. The greater the value placed on the protective, landscape and recreation functions of forests, the more acceptable have been the arguments to leave stands to age.

The former socialist countries recognized the need to utilize their wood resources fully, but the social system was not capable of achieving this target.

With regard to the other Central European countries, the prescribed rotations in Denmark are reported to be 50 to 150 years for pine depending on site quality, 40 to 120 years for spruce and 85 to 150 years for broadleaved trees, mainly beech. Relatively young classes predominate in the coniferous forests. As the growing stock becomes older, it is possible that the length of rotation may be increased. It seems at present that the economic profitability of wood production has a higher status as a criterion of successful forestry in Denmark than it has in the other parts of Central Europe.

4.4.4 Atlantic Europe

In Atlantic Europe, the original natural forests were predominated by oak, beech and other broadleaved trees. Only remnants of them now remain. Their value to the wood economy is small, but for landscape, environmental and recreation reasons their value is considerable in such a densely populated region.

During the two World Wars, the strategic value of forest resources was recognized. Large experiments of afforestation by transfers and exotic conifers were started after World War I in order to determine the most profitable tree species for re-establishing forests. Local Scots pine, transfers such as Norway spruce, European larch and black pine, and exotics such as Sitka spruce, lodgepole pine and Douglas fir were used in the experiments.

World War II intervened and 60% of coniferous and 40% of broadleaved forests were felled in the United Kingdom to support the war effort.

Net discounted revenue and the opportunity cost of invested capital of at least 5%, real value, were the criteria used when purchasing land for afforestation, choosing tree species, stand establishment and tending methods, as well as thinning and rotation regimes. Of the 5%, 1.5% was expected to be obtained from rising wood prices.

As a result of the economic analysis, Sitka spruce and lodgepole pine on drained peat lands proved to be the most profitable tree species. Douglas fir is valuable for the production of high-quality sawlogs. Conifers account for about 70% of the forests planted in the 1980s and the most valuable broadleaved trees, oak and beech, 9% and 4% respectively.

Mineral sites are often characterised by a layer of peat on the surface. Ploughing and fertilisation are therefore profitable land improvement measures.

The planting density is about 2300 per hectare, considerably less than in Central Europe but about equal to the density used in Northern Europe.

Relatively early and heavy thinnings increase the profitability of the stands. However, in the 1980s, because of the great risk of windthrow and high logging costs on the hills, 40% and 10% respectively of the Sitka spruce stands were prescribed to grow without thinnings and with somewhat shorter rotations than on other sites.

The most significant difference between the management regimes of Atlantic and Central European forestry concerns rotations. In Atlantic Europe, the prescribed rotation is 50 to 60 years for pine, approximately 50 years for other conifers and 120 to 150 years for oak and beech. In spite of these short rotations, the proportion of sawlogs was, at the end of the 1980s, about 60% of the coniferous removals. The quality of logs is obviously poorer than that of the logs grown by longer rotations, but the demand for them in the domestic market is sufficient.

The management regimes in Atlantic Europe are typical for conditions where the national aim is to build up forest resources by afforestation. Large investments should earn a reasonable return and the marginal rate of return of the growing stock should approximately equal the minimum net discounted revenue for a stand. It may be that when forestry is a going concern, the initial investments will be considered to be sunken costs and the rotations become longer than now prescribed.

4.4.5 Sub-Atlantic Europe

A great number of tree species, mostly of local provenance and some local transfers, is a typical feature of Sub-Atlantic European forests. Broadleaved trees predominate except in the Benelux countries where conifers have been favoured in afforestation activities. Afforestation has also doubled the proportion of coniferous stands to about 30% during the period of 1980 to 1990. Coniferous plantations have their largest extent in the southwest of France.

The proportion of coppice and coppice with standards varies from 9% of exploitable forest in the Netherlands to 50% in France. Their total area is 7 million ha. They mostly satisfy the traditional needs of the agricultural population. In spite of the low value of their removals there are institutional barriers to changing them into high forests. Fast growing poplars now account for about 300 000 ha.

The prescribed and actual management regimes in the eastern parts of the region on the slopes of the Jura and Vosges mountains, where spruce and fir are the main species, are similar to those in Central Europe. Prescribed rotation ages for coniferous, mainly pine, stands in the southwest of France are shorter than in the east.

In the Benelux countries the management regimes for thinnings and rotation ages lie between those of Central and Atlantic Europe. The aim is to increase the self-sufficiency of forest products by intensive forestry.

The most southernly parts of France are located in the Mediterranean climate where the density of forests is low and the proportion of scrub vegetation high. Wild fires, mostly ignited by man, are a great hazard, although most fires occur in scrub forests.

4.4.6 Pannonic Europe

Both broadleaved and mixed broadleaved-coniferous forests grow on the plains, hills and lower mountain slopes, while mixed coniferous and broadleaved forests grow on the Carpathian and Transylvanian mountains. Most of the original forests deteriorated because of exploitative logging and the utilization of forests by the large agrarian population.

Considerable forest improvement and afforestation, in which conifers have been favoured, has been accomplished during recent decades. Planted stands are now dominant in Hungarian production forests. Local tree species and transfers have been used in reforestation, and the only notable exotic is the false acacia; this now accounts for nearly 20% of the closed forests. Fast growing poplar hybrids are also grown.

Broadleaved stands predominate in the region. Coppice amounts to 29% of the exploitable forests in Hungary and 8% in Romania. Summer and evergreen oaks, beech and hornbeam grow on the hills. Because of the sharp continental climate, conifers grow best in the mountains and mixed forests on the mountain slopes. Spruce and fir are the most important species in the mountains.

Because of considerable demand for domestic wood, the management regimes in Hungary are efficient. The prescribed and actual rotation ages are 60 to 70 years for pine, 70 to 100 years for oaks, 100 to 120 years for beech, 30 to 35 or even 75 years for false acacia, 15 to 40 years for poplars and 30 to 75 years for coppice and coppice with standards.

The silvicultural regimes applied in the coniferous mountain forests in Romania follow the Central European forestry tradition. Both the density and rotation age of the stands are high. The prescribed rotation ages are 100 to 110 years for conifers and 100 to 120 years for broadleaved trees.

Spruce and fir forests may be facing the risk of density and age instability. Planted pine on the plains and low hills may be too far outside their optimum growing conditions for long term stability.

4.4.7 Mediterranean West

The tree-species composition of the Mediterranean West region is characterised by a great number of coniferous and broadleaved tree species, amongst them exotics such as eucalyptus and Radiata pine, as well as poplars. Closed stands of the original forests can be found only in the most inaccessible mountains. Coniferous forests account for 56% of the exploitable forests in Portugal and 47% in Spain; broadleaved high forests account for 26% and 40%, and coppice 12% and 13% respectively. Scrub vegetation covers millions of hectares.

Concerning management regimes, a sharp distinction should be made between forests in which wood production predominates and forests which are managed to produce wood, cork and mast. The multifunctional broadleaved forests are composed of stands of low density and high rotation age.

There are large areas reforested or afforested with coniferous and broadleaved trees in this region. A large proportion of these forests protect against erosion and the silting of waterways and electric power reservoirs. According to the Spanish Forest Statistics 1983, density of stands varies from 27% to 67% irrespective of species composition compared with the full density of 100. In Portugal, there is a long tradition of managed maritime pine plantations, while the oldest stand of eucalyptus was planted in 1907.

The prescribed rotation age is about 50 years for maritime pine and 60 to 100 years for other pines, 150 to 200 years for cork and other oaks, 8 to 15 years for eucalyptus and 8 to 20 years for poplars and coppice.

Although the production of industrial wood is rapidly increasing, the future of forestry is not quite clear. The controversy between (i) the traditional uses of forests and the growing of industrial wood and (ii) between the water consuming eucalyptus plantations and other users of water, together with the prospects, the risk factors of wild fires need to be solved if commercial forestry is to be fully accepted by society.

4.4.8 *Mediterranean Middle*

The great variety of soil parent materials and land relief from high mountains to coastal plains with corresponding differences in temperature, precipitation and wind conditions, has resulted in a wide range of vegetation types. Factors destroying the ancient forests and the unstable political history of the past thousand years, as well as a continuation of traditional uses of forests explain, at least partly, why the social basis for managed forestry is not the best possible. The current violent disintegration of the former Yugoslavia is eroding the results of sustainable forestry achieved so far.

Broadleaved stands predominate in the region. Coppice accounts for 55%, 38% and 39% of exploitable forests respectively in Italy, Yugoslavia and Albania. Pressures from a dense agricultural population restrict the managed production of industrial wood. Small isolated stands of the original forests can be found only on the slopes of the Italian Alps and in the Yugoslav mountains.

Poorly developed markets for industrial wood restrict the application of wood-production regimes, although there have been efforts to produce industrial wood in plantations of fast growing conifers, poplars and eucalyptus. Poor price developments for wood have eroded these efforts.

The forest management regimes used in the coniferous forests of northern Italy and Yugoslavia resemble those applied in Central European regions although rotation ages are shorter. Management regimes in the forests predominated by broadleaved trees are attempts to satisfy traditional and industrial demands for wood.

4.4.9 Mediterranean East

The tree-species composition, structure and quality of forests, management regimes, the history of forests as well as the current social conditions in this region resemble those of the other Mediterranean regions.

The Bulgarian management regimes are not dissimilar to those applied in Pannonic Europe. Pressure from agrarian populations restricts the efforts to improve the conditions for industrial wood production. In Greece and Turkey, the agrarian population continues to defend the traditional right to harvest wood for fuel and other household purposes in forests irrespective of ownership.

The proportion of degraded forests is high, e.g. more than a half of the forests in Turkey. Coppice accounts for 27%, 48% and 30% of exploitable forests in Bulgaria, Greece and Turkey, respectively.

In spite of the efforts to improve forests, their degradation continues. One consequence of this may be the considerable decrease of removals reported to FAO in the 1980s.

4.5 Stability of the current forests

To achieve stability is a general target of forest policy. The concept and prerequisites of it are, however, not always defined explicitly. From the point of view of wood production, stability, or sustainability, is attained when the volume of harvested wood does not exceed the volume of the increment or the sustainable allowable cut estimated by long-term production scenarios. In the current stage of European forestry, fellings have been and will continue to be so much smaller than net increment that the sustainability of production is guaranteed for the foreseeable future.

Deteriorating environmental conditions have called for a global strategy for sustainable development. In this strategy, the concept of ecological sustainability has general importance. The primary criterion of ecological sustainability is the preservation of the process-functional entity of the forest ecosystem as a hierarchical body of interacting plants, animals and micro-organisms in an environment of physical and chemical processes. The richness of an ecosystem is indicated by its biodiversity, measured as the number of species in the system.

From the point of view of forest policy, however, an essential quality of a stable forest is the sustainability of the functional services of the forest, demanded and valued by people and national economies (Fig. 4.1).

Unhealthy and dying trees are considered to be symptoms of decreasing stability caused by storms, insects, fungal diseases, accumulated pollution, etc. Decreasing stability is relieved by sanitary and emergency cuttings in situations when dying and dead trees cannot be harvested by regular thinning and final cuttings.

4.5.1   *Genetic stability*

A stand or forest is considered to be genetically stable when the constituent species successfully endure and adapt to the dynamic environmental factors under which they grow. The genetic qualities of trees on a particular site can be classified as locals, transfers, exotics, or culturals, i.e. hybrids or similars.

Locals are those tree species which have grown in a specific locality and environment for thousands of years. Their genetic structure is fully adapted to the climate and its variations, soils and water regimes. The current trees are grown from the seeds or sprouts of the locals.

Transfers are trees grown from seeds or crafts of an origin grown at a distance relatively far from the locality, e.g. from a neighbouring forest zone on the same continent.

Exotics are trees moved from another continent or from environmental conditions which differ markedly from those where the species is planted. In this study the tree species moved from other continents are called exotics and those moved within Europe are called transfers.

Hybrids of different species of a genus and seedlings micropropagated from tree tissues can be called cultures. Most of them, such as poplars, are grown in short-rotation plantations where the intensity of management is closer to agriculture than forestry.

Lodgepole pine is the only exotic tree grown in the production forests in Northern Europe. It accounts for about 330 000 ha in northern Sweden. Stands have grown well during their young stages. Fungal diseases have occurred, however, and planting has been restricted while more experience is gained. The ability of lodgepole pine to regenerate naturally is probably poorer than the ability of local trees.

On the whole, the genetic stability of the forests in Northern Europe is good because almost all tree species are locals. Fungal diseases occur more frequently in those seedling stands established from seeds of localities more than about 150 km south or north. The guidelines controlling the transference of seeds have been amended accordingly.

There have been large-scale experiments with exotics, transfers and larch hybrids in the United Kingdom since World War I. These experiments have indicated that the yield of exotics and transfers can be estimated reliably when the experiments have lasted for at least half the rotation. For example, many species can grow quickly in their seedling stage but their growth declines later.

Sitka spruce, and lodgepole pine on peat sites, have proved to be the most profitable trees. The bulk of the current production plantations is composed of these two exotics. They have been a great success and they are the resource base for the expanding forest industries. Young age classes (under 40 years) predominate at the moment. When the even age structure of the target forest is reached, the annual increment may well equal the climatic potential of 8 to 10 m³/ha.

The exotic conifers of Atlantic Europe (Fig. 4.1) grow in the zone where oak, beech, ash and other broadleaved trees are the natural climatic climax species. Stormdamage is the greatest risk threatening the stability of the coniferous stands. In order to minimize such damage, the planned thinning and rotation regimes are changed on hill slopes and sites with soft soil-parent materials so that a part of the stands is grown unthinned on a relatively short rotation. Sitka spruce in Europe is also sensitive to bark beetles. On the other hand, Sitka spruce regenerates well naturally, at least on some sites.

In Central, Alpic and Pannonic Europe there are relatively few exotics in the production forests. False acacia in Hungary and Douglas fir in some Central European countries are examples. The stability of these stands seems to be good. The planting of Douglas fir may increase because of its good quality for sawlogs.

Eucalyptus is the most important exotic in Mediterranean West and to minor extent also in other Mediterranean regions. It is a growing resource for the pulp and paper industries. Plantations have grown well and they can be regenerated by sprouts for a limited number of rotations.

The greatest problem with eucalyptus is their great need for water in the areas of low precipitation; frost also limits their range. The shortage of water limits the yield of wood, while the high transpiration of plantations is considered to be harmful for the water regime of adjacent areas. The future yield of plantations will obviously be smaller than that expected when the planting was started, but whatever the future of the Iberian eucalyptus forests, they are man-made artefacts having to be maintained by intensive silvicultural measures.

Radiata pine is a fast growing and productive exotic in the maritime parts of Iberia. It covers about 250 000 ha in Spain. The stability of the stands, as man-made artefacts, seems to be good.

Although relatively little discussed, the greatest issue concerning the stability question is the large coniferous forests growing in Central European localities where oaks and beech are the natural climax species. It is estimated that on 80% to 90% of the sites of the ancient natural forests oaks, beech and other broadleaved trees predominated, whereas there is now a 60 to 70% predominance of coniferous trees.

The origin of the transfer seeds used in the large afforestations of the 18th and 19th centuries is not accurately known. The seeds used in the deciduous broadleaved zone proper can be only from the localities of the ecotonal mixed and Sub-Alpine zones. At the same time, at least part of the seeds used in the planted coniferous stands on hills and mountains were collected from lower elevations.

The bulk of the spruce and fir stands, as well as the stands of pine outside the outwash sands, in the deciduous broadleaved zone are plagio-climaxes maintained by silvicultural measures. If the silvicultural measures were abandoned and the coniferous forests left to natural forces, broadleaved trees would begin successions to re-establish the original climatic climaxes.

A symptom of the genetic instability of these forests is the occurrence of extensive storm damage. In particular, the dense maturing and mature stands of tall spruces cannot endure severe storms. Genetic instability may also be a co-factor causing the much discussed forest decline

(*Waldsterben*).

From the point of view of genetic stability, cultures such as poplars are not problematic. Fast growing and intensively managed by short rotations, they are more like agricultural than forestry crops.

4.5.2 Density and age stability

In multi-functional forests which satisfy the needs of both people and economies, forest decline is considered to symptomize instability. Critical stages in forest decline are the increasing density, age and volume of the growing stock. When these stages are exceeded, a part of the expected benefits from the forest is lost. For example, postponing thinning causes natural losses and lowers the value of trees in fellings. Even if the prescribed thinnings are carried out, increasing age and volume decreases the marginal rate of return on the growing stock. At a certain point the rate of return is zero and after this point it is negative. If this development continues, the growing stock begins to degrade. The natural forces such as storms, fungal diseases and insects cause more and more losses. Further, in dense and heavily stocked old stands the transpiration of water is high in relation to their net biological production. In growing seasons drier than average, trees suffer from a shortage of water, and lose their vigour and a part of their needles or leaves.

Information concerning the age, density and growing stock volume of forests in European countries is insufficient for accurate analysis of the current situation. However, it is obvious that a great many stands have reached the stage where self-thinning decreases revenues. Stands are reaching the stage where the marginal rate of return is zero or negative, and where the deterioration of growing stock has begun.

The idea of age instability can be illustrated by the critical stages of the growing stock of forests predominated by Norway spruce on an area of 270 000 ha in southern Finland (Fig. 4.3). Alternative growing stocks are presented as a function of rotation age. The influence of the thinning regime is eliminated by applying the current regime for all stages of growing stock development. In the fully regulated target forest the rotation age is 80 years, the mean volume 132 m^3/ha, and the marginal rate of return of the growing stock is 3 to 4%.

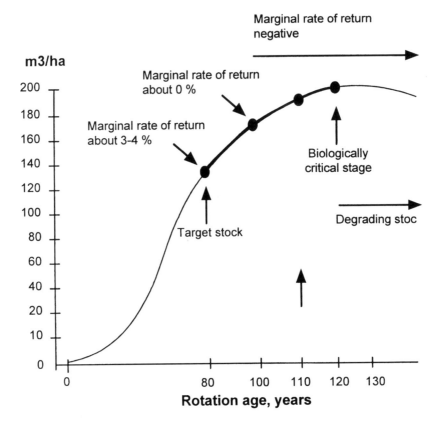

Fig. 4.3   *Critical stages in the growing stock of a forest predominated by Norway spruce in southern Finland as the function of rotation age*

The rotation age of the current growing stock has increased and is now about 110 years, and the mean volume is 185 m³/ha. The longer rotation age has increased fungal diseases and decay losses to such an extent that the value of the crop trees is decreasing. The marginal rate of return of the growing stock is negative.

In the forest yielding the greatest income per hectare, the rotation age is about 100 years, the mean volume 170 m³/ha and the marginal rate of return of the growing stock is about zero. This stage of development is critical from the point of view of economic profitability.

When the rotation age increases to about 120 years and the mean volume to about 200 m³/ha, the forest has reached the biologically critical stage. Trees are dying because of old age. The losses caused by fungal diseases are large. Consequently, about 20% of the high value sawlog volume will be lost to low value pulp wood or logging losses.

The growing stock reaches the degrading stage at a rotation age of about 130 years. The volume of sound trees decreases, and value losses undermine the profitability of forestry, while environmental benefits also deteriorate.

4.5.3 Health stability

Public debate in Europe in the 1980s has addressed the concepts of 'new forest damages' and 'forest decline' more than any other forestry issue. It has continued in the 1990s, although in somewhat abated form.

The harmful effects of the emissions from the power and metallurgic industries have been recognized and studied since the 19th century. The current popular interest in 'forest decline' is based on its status in the mass media and as a means of getting publicity and financing for research projects to study the health of forests. The inconsistency of the concept of what is a healthy or an unhealthy forest has also fuelled the discussion.

In the first place, the concepts of forest, tree species and tree individuals have been confused. A forest is composed of stands distinguished from each other on the basis of site quality, species composition, age of trees, density of stand and other relevant characteristics. Stands are easily recognisable under the conditions of intensive silvicultural management, especially in plantation forestry. In extensive forestry there are also natural untended stands which differ as ecological entities from managed stands. In Northern Europe, where there are still virgin forests, stands are large and they include a variety of tree species, age and density composition.

A stand is composed of tree species and individuals. In natural stands, a species may decline as another species gains more dominance. In the competition between individuals for nutrients and space, some individuals lose their vigour and foliage and finally die, while stronger individuals survive and gain more dominance. This is an essential feature of a healthy stand and a healthy forest and not a symptom of unhealthiness. These phenomena come under the headings of genetic, age and density stability.

An unhealthy and damaged individual, species, stand or forest is caused by direct and indirect influences of man or environmental factors. Anthropogenic emissions from production, traffic and consumption are indisputable health factors. Direct fumes injure the tissues of trees, leading to defoliation and finally death. Such damage is worst adjacent to metallurgical and power industries where they have destroyed forests over limited areas and changed them into treeless plant communities or even into more or less barren ground.

Disturbing effects are also caused by emissions transported by wind. They precipitate over large areas and distances of hundreds or even thousands of kilometres from their sources. Because of the complex and not thoroughly understood nature of their effects, most of the current opinions concerning the effect of emissions on the health of forests should be considered as hypotheses. It has been observed, for example, that emissions have changed the species composition of plant communities, e.g. those of lichens living on trees. It can therefore be hypothesized that emission effects may also change the course of plant community successions and the species composition of climatic climaxes.

The precipitation of sulphur dioxide, nitrogen oxides and hydrochloric and hydrofluoric acids as acid rain cause direct damage to plants, as well as acidifying soils, therefore possibly disturbing nutrition regimes. It is to be expected that the biological activity and fertility of soils may decrease if the precipitation of emissions continues at the current rate.

As noted, the observed effects of emissions are complex. For example, increasing nitrogen initially improves site quality and increases biological production, especially in boreal ecosystems where active nitrogen is generally the limiting nutrient. Increasing the carbon dioxide content of the atmosphere also increases biological production. This may partly explain the increasing increment of stem wood in European forests which has exceeded earlier forecasts.

Information concerning the severe damage to forests caused by emissions is inconsistent. The decline area may be as little as 0.5% of the forest area, located mostly in Poland, Czechoslovakia and Germany. There are no reliable quantitative estimates of the role of climatic factors such as shortage of water, warm winter winds which start transpiration while ground water is still frozen, exceptionally cold air and desiccating winds, as well as weather conditions which disturb trees preparing for winter dormancy. Genetic, age and density instability and biologically critical stages of the growing stock also act as potential decline factors.

If the increasing increment of the growing stock volume is considered as demonstrating the health of forests, emissions have not had a deteriorative effect so far. Nonetheless, whatever the cause and effect relationships in forest health, the emissions of harmful residues should be decreased to the largest extent possible. Although these measures are outside the traditional activities of forestry, those responsible for the health of forests should increase their activities to affect a decrease in emissions.

Considerable damage caused by wild fires occurs in dry periods, especially in the Mediterranean region. Many such fires are caused by careless or deliberate human action. A relatively large part of burned-over wooded land grows scrub vegetation, but considerable damage is also caused in stocked forest and even in young planted stands. The only means effectively to prevent the fire hazard is education to encourage people to see forests as a part of the national or local wealth. Without such a feeling of public responsibility, even the most efficient fire fighting measures are useless.

Browsing damage to young stands and also to old trees is of anthropogenic origin. The stock of game animals is excessive because of hunting interests and because the influence of carnivorous predators has been eliminated. If the income from hunting rights, meat, skins and trophies is large enough to compensate for the resultant losses in wood production, game animals become a part of the integrated forest profitability. On the other hand, if the damage caused to wood production is too great, the only solution is to reduce the game stocks and to seek a balance between the interests of wood production and hunting.

Insect damage and fungal diseases cause considerable losses to wood crops. They are nature's means to eliminate biologically critical stages of the growing stock. These damages are partly caused by management regimes which maintain over-mature, over-dense and degrading stands.

Insect damage can be prevented to a certain extent by encouraging stocks of insectivores, but under favourable climatic conditions, the insect stocks can increase to such an extent that chemical and biological preventives must be used.

The damage problem is complicated where transfers and exotics are more sensitive to local insects and fungal diseases than the local trees, or whose damage factors have been imported from other regions and continents, e.g. with imported roundwood, chips and mechanically processed wood products. Testing provenances, controlling the trade of seeds, seedlings and forest products, and developing forest hygienic measures are means to prevent such damage.

4.6 *Stages of economic development*

Stages of economic development are illustrated by countries' gross domestic product (GDP), and by their consumption of major processed wood products, expressed per capita, in Table 4.2 and Fig. 4.4. The size of population is also presented in Table 4.2. The information is taken from the UN-ECE/FAO Assessments 1986, Volume II and 1992, Volume I. The estimates may differ from other statistics, but they provide a sufficient basis for general observations.

From the point of view of wood production, the quality of forests is poorest where the growth of population and the proportion of the farming population are greatest. In the Mediterranean region these characteristics are often combined with a low domestic product per capita. In the former socialist Eastern European countries, the domestic product is low but the quality of forests is relatively good in the countries which are within or close to the Central European silvicultural heritage.

The worst obstacles restricting efforts to improve the wood-production function of forests are a low material standard of living and a high rate of population growth combined with a considerable agrarian population. The heritage of multi-product management in which the material benefits are composed of wood, pasturing, mast, cork, etc., is also an obstacle to wood production proper.

From Fig. 4.4, it can be seen that the region-wise consumption of sawnwood is in a weak positive correlation with GDP. Sawnwood consumption is relatively higher in those countries where the forest area per capita is high and where wood has traditionally been an important material for buildings, etc. Increasing urbanisation also decreases the use of sawnwood for these purposes. Further, the general goal of accumulating growing stock restricts the use of domestic sawnwood in Central and Alpic Europe.

The positive correlation between GDP and the use of wood-based panels is greater than that for sawnwood. The higher the technical level of industries and the greater the forest resources, the greater is the use of wood-based panels. The national ability to adapt new building materials and access to large-scale wood-based panel industries are factors favouring the use of panels.

There is a high positive correlation between GDP and the use of paper and paperboard. The use of paper and paperboard per capita can be used as an indicator of material living standards. Rich countries with small forest resources import the amounts needed to supplement their own domestic production.

Table 4.2   *Population in 1980 and 1990 and its change, Gross Domestic Products (GDP) per capita, and the consumption of sawnwood (SW), wood based panels (WBP) and paper and paper board (P+PB) per capita in 1980*

Country or country group	Population, million		Change 1980-90	GDP/cap 1980
	1980	1990	%	1000 USD
Finland	4.8	5.0	4.20	9.30
Norway	4.1	4.2	2.40	14.00
Sweden	8.3	8.6	3.60	13.60
Denmark	5.1	5.1	0.00	11.00
Germany W.	61.7	63.2	2.40	13.40
Germany E.	16.7	16.7	0.00	5.70
Poland	35.8	38.2	6.70	1.80
Czechoslovakia	15.3	15.7	2.60	5.90
Ireland	3.4	3.5	2.90	4.60
United Kingdom	55.7	57.4	3.10	9.30
Netherlands	14.2	14.9	4.90	11.20
Belgium	9.8	9.8	0.00	12.10
Luxembourg	0.4	0.4	0.00	
France	53.8	56.4	4.80	12.10
Austria	7.5	7.7	2.70	10.30
Switzerland	6.4	6.7	4.70	15.90
Hungary	10.7	10.6	-0.90	2.30
Romania	22.2	23.2	4.50	1.90
Portugal	9.7	10.5	8.20	2.30
Spain	37.5	39.0	4.00	5.40
Italy	56.2	57.7	2.70	7.00
Yugoslavia	22.3	23.8	6.70	2.60
Albania	2.7	3.3	22.20	(1.0)
Bulgaria	9.0	9.0	0.00	2.70
Greece	9.6	10.1	5.20	3.70
Turkey	44.5	58.7	31.90	1.30
Cyprus				
Israel				
Northern	17.2	17.8	3.50	12.50
Central	134.6	138.9	3.20	8.40
Atlantic	59.1	60.9	3.00	9.00
Sub-Atlantic	78.2	81.5	4.20	11.90
Alpic	13.9	14.4	3.60	12.90
Pannonic	32.9	33.8	2.70	2.00
Mediterranean W.	47.2	49.5	4.90	4.80
Mediterranean M.	81.2	84.8	4.40	5.60
Mediterranean E.	63.1	77.8	23.30	1.90
Europe	527.4	559.4	6.10	7.30

Table 4.2   Continued

Country or country group	Consumption per capita in 1980		
	SW dm3	WBP dm3	P+PB kg
Finland	646	145	237
Norway	602	155	127
Sweden	616	148	200
Denmark	377	133	149
Germany W.	234	130	156
Germany E.	223	85	83
Poland	189	58	38
Czechoslovakia	253	79	70
Ireland	187	51	80
United Kingdom	152	61	129
Netherlands	222	87	155
Belgium	202	84	136
Luxembourg			
France	218	62	115
Austria	392	93	122
Switzerland	331	103	159
Hungary	192	44	61
Romania	164	54	31
Portugal	124	36	45
Spain	92	38	71
Italy	142	61	92
Yugoslavia	157	53	48
Albania	74	4	5
Bulgaria	166	54	54
Greece	78	47	46
Turkey	104	11	12
Cyprus			
Israel			
Northern	621	149	193
Central	228	99	105
Atlantic	154	60	126
Sub-Atlantic	217	69	125
Alpic	364	98	139
Pannonic	173	51	41
Mediterranean W.	99	38	66
Mediterranean M.	144	57	77
Mediterranean E.	109	23	23
Europe	192	68	92

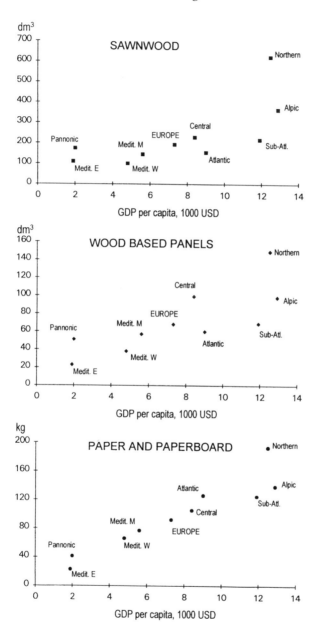

Fig. 4.4   *Consumption of major forest products per capita, by country groups in Europe*

An increasing GDP per capita has the greatest effect on increases in the consumption of paper and paperboard, but it has relatively less effect on the consumption of wood-based panels and least of all on the consumption of sawnwood. Efforts to increase the material living standard in those countries where the current standard is low will increase the total use of wood and wood-based products.

4.7 *Multi-functional management*

Forests have always played a multi-functional role in the history of mankind. Initially, populations were small and the forests seemingly endless. In his use of foliage, edible plants, game and wood, man lived as a part of the forest ecosystem. In historical times, populations have increased and forests have given way to agricultural land to such an extent that forest resources are increasingly scarce. Further, the development of forestry was initially to manage royal hunting estates and then later to produce industrial wood on a sustained basis. The two management functions were not considered to be mutually exclusive.

The modern concept of multiple-use forests is a product of welfare societies, where the proportion of people working in the primary sector has become very small. The remaining sectors of the populace have therefore become estranged from their material dependence on forests and now demand leisure-related multiple benefits from forests: benefits which were formerly self-evident. Multi-functional forestry has therefore become a fashionable concept in modern welfare societies, in spite of the fact that the contents of the concept are obscure and often contradictory.

This is demonstrated by the efforts to study and define the relative importance of forest benefits: '...the role of non-wood benefits in the forests of the ECE region is incompletely described, although they are reckoned to be of significant importance...The value of goodwill obtained through adopting multiple-use concepts could easily compensate for any loss in wood production' (*The Forest Resources of the Temperate Zones*, Volume II, p. 227). In other words, even if the multiple-use concept can only be incompletely defined, it will be practised in order to achieve the goodwill of the general public.

The situation can be illustrated by some answers to the inquiry of the 1990 Resource Assessment.

The reported importance of the wood production function varies irrespective of the forest area per capita and the importance of forestry in the national economy. Using an index of 100 to represent singular importance, and zero (0) to represent no importance at all, the importance of the wood production function is reported as 47 in the public forests of Sweden, 56 in Denmark, 98 in Germany West, 52 in Poland, 80 in Czechoslovakia and 100 in Ireland.

The importance of the protective function of public forests is 0 in Ireland and 10 in the United Kingdom, 9 in Germany West and 1 in Germany East. The importance of water management function of forests varies randomly. That of the grazing function in public forests is reported as 81 in Finland and 0 in Sweden, in spite of the fact that the reindeer stock is more or less equal in both of these countries.

The importance of hunting is reported as very small in Northern Europe where hunting is the local public right in the state forests and a right tied to forest ownership, whereas its importance is great in those countries where hunting is a luxury of relatively few. The importance of the nature conservation and recreation functions has no correlation with the forest area per capita or with the areas where nature conservation proper is mostly needed.

The answers are partly contradictory, a problem caused by the specifications of the inquiry. A high importance of wood production is specified as follows: Annual cut exceeds 3 m^3/ha per annum on average over a long period (50-100 years)'. In the boreal coniferous zone there are large areas where the actual and potential yield is less than 3 m^3/ha per annum but still the wood production function is the most important. The importance of grazing is also tied to the amount of fodder calories produced during the vegetative period, while the potential production is different under different climatic and soil fertility regimes.

In spite of these critical remarks there is a need to recognize and define the multiple functions of forests, to estimate the importance of each function in different climatic and cultural regimes with due regard to the traditional, current and future land uses, and to recognize the interdependence of the functions. Here, it is only possible to illustrate the principles on which the problem could be tackled and on which a sensible multi-functional management of forests might be based. As a frame of reference the functions are as follows.

1.   The commodity function
        - wood production function
        - other commodity functions: production of cork, edible seeds,

mast and other fodder for domestic and wild animals, berries, mushrooms, medical and decorative plants, game animals, etc.

2.   The protective function. Forests are employed to control erosion, landslides, avalanches, floods, silting, strong and desiccating winds, noise, emissions, etc. Forests in need of regulated management measures in order to prevent the advancement of tundra and steppe into the area of closed forest can also be classified under this function. They are called protection forests in Northern Europe.

3.   The social function. Forests here provide a healthy living and working environment with scenic and recreational values. Recreation includes leisure hunting and fishing and the gathering of berries and mushrooms.

4.   The cultural function. Forests provide aesthetic and symbolic values and preserve nature reserves and historical monuments.

The benefits from the commodity functions and also from part of social function can be valued in monetary terms. The non-material values can only be based on an agreed valuation system. A comparative sequence of the valuations of the material and non-material benefits needs to be constructed to assist the planning of balanced multiple function management regimes.

Under successful multi-functional management, all the benefits are specified as objectively as possible and they are integrated to give the greatest total benefit to society. All benefits, except those provided by strict nature reserves and virgin forests, can be provided only by proper management regimes and treatment measures. In all treatment measures wood is either the main or a side product. The value of other benefits is reduced or lost without a properly managed wood production function.

A forest in which the growing stock is degrading because of high density or age does not provide the best recreation benefits. Similarly, the production of cork and mast is greatest in forests where the spacing and crowns of trees maintain suitable production conditions.

The costs of multi-functional management can be met by selling wood and other commodities as well as the rights to enter the forest or to hunt and fish. The costs can also be subsidised by public money in welfare societies which have a relatively small area of forest per capita. In those countries where the forest area per capita is large, the income of the wood production function is the major source of finance for the costs of many other benefits. Profitable forestry is the only sound basis for good multi-functional management.

Maintaining the biodiversity of forests is considered to require certain areas of natural reserves left outside all production measures. As far as this issue is concerned, the countrywise conditions cannot be analysed on the basis of the current information. The concepts of national parks, other nature protection forests and nature reserves differ in different countries. For instance, national parks are strict natural reserves in Nordic Europe, but in many other countries human activities and measures are practised in national parks and nature protection forests. The forest statistics should be improved in such a way that the degree of protection can be analysed on the area of so-called nature protection forests.

The increasing pressure of national and international nature conservation organizations and green movements is confusing attitudes towards forestry and changing management regimes. The basic idea of sustainable development is accepted by the majority of people, not least by those responsible for forests and forestry. The confrontation between commercial and environmental interests is the result of the green movement, some of whose members mix conscious and unconscious motives and even sometimes adopt antagonistic attitudes towards material production and social institutions. Their message is often based on unrealistic and unscientific arguments. If it were generally accepted, forestry and the forest industries would be in great difficulties, also in those countries whose national economies depend upon the export of forest products.

For instance, parts of the green movement are against the harvesting of wood from the existing virgin forests. If this concept were widely applied, it would decrease the production of wood not only in tropical rainforests but also in North America, Europe and Asia. It also opposes the production of wood by planted forests, especially those of transfer and exotic tree species. If this were generally accepted, the exploitation of the existing virgin forests would increase, since the world population must satisfy its need for wood in one way or another.

Some of the green movement is campaigning for management regimes which apply selection cuttings and which have been shown to erode the quality of forests, create inefficient regeneration methods and accumulate decaying wood in the forest. If accepted, this would increase the price of wood and so increase the pressure to exploit the existing virgin forests.

Consequently, certain segments of the green movement are working against sustainable development as defined in section 4.8, and even supporting the interests of the oil and plastic industries which are actually serious threats to sustainable development. Therefore information about the role of forestry and functional management regimes should be focused on those qualities of forests, forestry and uses of wood which support sustainable development - a position with which the green movement should have no quarrel.

4.8   Forestry supporting sustainable development

Sustainable development has been variously defined by the scientific community as an effort to preserve the benefits of renewable natural resources for mankind now and in the future. With regard to forests and forestry, the global targets are to stop the decrease in the area of forests in developing countries where the density and growth of the population are high, to maintain the functioning of forest ecosystems, to preserve the biodiversity of forests, and to decrease the net emission of carbon dioxide into the atmosphere. The key instruments are to increase the area and biomass of forests as well as increase the use of wood as a sustainably managed renewable resource.

The primary biological process of photosynthesis, and the decomposition of litter and dead trees is a process involving the throughflow of solar energy and the cycling of carbon from the air to trees and back to the air. There is no net emission of carbon. Trees grow, die and decompose as a part of the natural balance of the carbon cycle.

Tree biomass is thus a huge store of energy and carbon. In one cubic metre of tree biomass there is approximately 210 kg of carbon. The energy value of one cubic metre of tree biomass with 20% water content equals the energy of approximately 185 equivalent kg of oil.

On the other hand, the decrease in the global biomass of trees by human interference leads to a net emission of carbon into atmosphere. This net emission has been estimated to be 2 gigatons per year, while the net emission of carbon from burnt fossil fuels is 6 gigatons. Conversely, increasing the tree biomass fixes carbon and decreases its net emission.

Neither the wood consumed by man, as commodities (including the waste wood from their manufacture) or burned for energy, create a net emission of carbon with respect to the energy and carbon balance of the forest biomass.

Consequently, the more wood that is used in the process of sustainable forest management and the more fossil fuels and energy intensive products such as metals, concrete and plastics are replaced by wood, the greater the net decrease in the net emission of carbon into the atmosphere.

Additionally, the residues of forest products can be used as fuel and wood in durable commodities such as buildings, furniture, books, etc., or consumption residues can be dumped into anaerobic conditions, which all decrease the net emission of carbon.

An increase in the forest biomass due to afforestation and improved management regimes would increase the net sink of carbon from the atmosphere. The sustainable growth and use of wood, as well as increases in tree biomass are therefore effective ways to reduce the greenhouse effect and to support sustainable development.

The increasing growing stock and its increment have been in harmony with sustainable development so far. On the other hand, if the growing stock is left unmanaged to become both denser and older and to reach degrading stages over large areas, energy and carbon are lost as decay products and cannot be used to decrease the use of fossil materials.

The idea of sustainable development combines the dual target of ecological stability and a stable and increasing production of wood. Forests are ecologically stable when the natural functioning of the ecosystem continues. Another feature of a stable forest is its great biodiversity.

The functioning of forest ecosystems can be sustained in forests managed on relatively long rotation ages. The harmful effects of emissions and their precipitation are the only threats to the continued functioning of forest ecosystems (see section 4.5.3).

In the case of biodiversity, the abundance of species has an intrinsic value. Their genetic inheritance is also a natural resource. A large number of species is also considered to be a part of the basis of functional stability.

The dynamic equilibrium of species composition is an essential feature of biodiversity. The numbers of species increase and decrease. Some species may disappear from an ecosystem while new species may appear. Genetic development continues and follows geological time spans. Exceptions occur as catastrophic events such as volcanic eruptions and rapid climatic changes.

Further, man's activities are not ethically sustainable if they jeopardise the existence of any species in the interest of short-term profit. Without this moral code, the consequences of the current development trends will erode the basis of life for mankind.

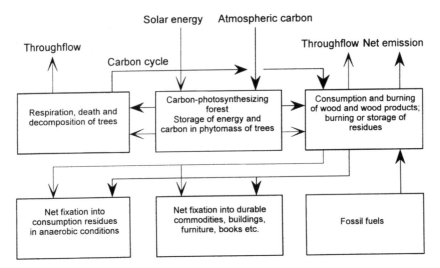

Fig. 4.5   *Throughflow of solar energy (—) and carbon cycle (-) in the forest, the atmosphere, the consumption of wood and its products, wood residues, and the consumption of fossil fuels as sources of net emissions of carbon*

4.9 Conclusions to Chapter 4

1. The qualities, development and utilization of forests should be studied inside the frame of reference of serial successions and climatic climaxes of plant communities by ecological forest zones, of the history of forests, types of economies using and managing forests, and the constraints of wood production.

2. Ecological forest zones are to a great extent regions inside which the quality, management regimes and uses of forests as well as the stage of economic development are also relatively homogenous.

3. The history of forests under human influence and the traditional uses of wood and other benefits go a long way to explain the current forestry conditions and determine which measures for developing forestry are realistic and successful.

4.   Economic criteria based on the opportunity cost of capital can be used for estimating the relative profitability of management regimes applied in regeneration, afforestation and forest improvement projects, especially in those financed by sources other than incomes from the selling of the existing mature wood crops. In commercial forestry, the net annual income and marginal rate of return of growing-stock capital are the more realistic indicators of economic success.

5.   The rotation age of fast growing poplars and eucalyptus is so short that criteria based on the opportunity cost of capital can be used as indicators of economic success.

6.   Where the basic investments in forestry are several decades old, the investments can be considered sunken costs. Consequently, they are of little or no value in defining the management regimes of the existing forests. In welfare societies where people are estranged from primary production and have plenty of leisure, the non-material benefits of forests often predominate management regimes with a consequent decrease in the profitability of wood production.

7.   In management regimes which apply long rotation ages, the sunken costs of forestry have little or no effect on wood prices which are determined by demand and supply in domestic and international markets.

8.   The primary features of management regimes by European regions are as follows:
     -   **Northern Europe.** Prescribed management regimes are based on economic criteria. With fellings much smaller than net increment, these regimes can no longer be followed. The marginal rate of return of the growing stock is decreasing and it is negative in large areas.
     -   **Central and Alpic Europe.** The present management of forests is characterized by high age and a large growing stock volume per hectare. The net income from wood production has been decreasing and the marginal rate of return is mostly negative. In many cases, the costs of forestry have to be subsidized.
     -   **Atlantic Europe.** Most of the coniferous forests have been established by afforestation. The bulk of the stands are at

thinning stages. Prescribed rotation ages are about half those applied in Central and Alpic Europe.

- **Sub-Atlantic Europe**. Management regimes are either the same as in Central Europe or rotation ages are between those of Atlantic and Central Europe. Low-quality coppice, which may have an environmental value, accounts for large areas.
- **Pannonic Europe**. Effective management regimes predominate in the cases of regeneration and afforested stands. Regimes of coniferous forests in the mountains are similar to those applied in Central and Alpic Europe.
- **Mediterranean West**. Management regimes are a mixture of those applied in fast growing plantations and those applied for other forest types. There are well-managed long-rotation forests, multi-benefit forests producing wood, cork and mast, and degraded low-density forests.
- **Mediterranean Middle and East**. Management regimes of the coniferous forests in the southern slopes of the Alps in the Yugoslavian mountains resemble those in Central and Alpic Europe. A great part of the other forests are managed on a multi-benefit basis or they are in various stages of degradation. There are projects to improve existing forests and to establish fast growing production forests.

9.   **As a whole**, the biological stability of European forests is satisfactory. The genetic qualities of trees transferred from other localities and exotics, as well as stands of high density and age cause instability problems. The greatest threat to stability is stand age which in many cases is approaching the stage at which degradation sets in.

10.   Health stability is endangered by the deposition of emissions from production, traffic and consumption. Direct fumes from power mining and metallurgical industries have reduced the health of trees and killed them in limited areas around emission sources. The harmful effects of acid rain are often confused with biological instability factors. Consequently, the exact role of acid rain has not yet been determined.

11.   From the point of view of wood production, the quality of forests is poorest in those regions where the growth of population and the proportion of the agrarian population are highest, often

combined with low Gross Domestic Product per capita. Under these social conditions it is difficult, if not impossible, to introduce effective management regimes for producing industrial wood.

12.   The consumption of paper and paperboard increases in close correlation with an increase in Gross Domestic Product per capita. Relatively less effected by a GDP increase is consumption of wood-based panels, and least of all that of sawnwood. Increasing material living standards in those regions where the current standard is low will markedly increase the consumption of industrial wood and wood-based products.

13.   Forests have always had a multi-benefit role. The current concept of multiple-use forest derives from welfare societies and is as such an artifact. Its concept and content are poorly defined. This is demonstrated by the contradictory and inconsistent answers to the UN-ECE/FAO 1990 Resource Assessment inquiry concerning the benefits and functions of forests.

14.   In order to use the concept of multi-benefit successfully in the development of management regimes, the individual interconnected functions should be defined with regard to their material and non-material benefits. Benefits which can be valued in monetary terms and those immaterial benefits, which can be subject to an agreed valuation system, should be separated from each other. On the basis of these analyses, a comparative sequence of valuations of all benefits could be developed for balanced multi-benefit management regimes.

15.   Forest benefits should be classified into (i) those which are financed by the incomes of wood production; (ii) those benefits other than wood which have market value; and (iii) those which are financed partly or totally by public money. The number of non-material benefits which have market value should be increased.

16.   In order to adjust the conflict between wood production interests and nature conservation organizations and green movements, forest information should be focused on those forestry measures and wood uses which maintain and create biodiversity and support sustainable development.

17.   Regeneration areas, successions of ground vegetation, boundaries
      between stands, stands of rare tree species and exotics, etc., should
      be recognized as cultural biotypes maintained by forest
      management, which supplement the range of biodiversity.

18.   The qualities of forestry which support sustainable development
      are as follows:
      - Sustainable and progressive forestry is based on the
        renewability of forest resources and a total of wood assortments
        which does not exceed the increment of a growing stock
        sustained by silvicultural measures, forest improvements and
        afforestation.
      - A primary target of sustainable forestry is carbon fixation in
        order to decrease the net emission of carbon dioxide into the
        atmosphere.

19.   Preserving the functioning and biodiversity of forest ecosystems
      is another target of sustainable development. It is attained by
      maintaining the genetic, density, age and health stability of forests,
      by applying management and silvicultural regimes which preserve
      and create cultural biotopes and biotopes of endangered species,
      and establishing strict nature reserves.

20.   Traditionally, man has been responsible only to himself and has
      satisfied his needs by using natural resources without considering
      their sustainability. Today there is a need for an environmental
      ethic to cover the whole biosphere to advance the survival of all
      plant and animal species. Current economic activities are not
      ethically sustainable if they jeopardize the existence of any species
      because of short-term profit interests.

REFERENCES

Dengler, A. (1980). *Waldbau*. 1. Band, 5. Auflage, neu bearbeitet von Ernst Röhring. Hamburg: Verlag Paul Carey.

Duinker, P. N. (1990). *Biota: Forest decline. Towards Ecological Sustainability in Europe*, ed. A. M. Solomon and L. Kauppi, pp. 115-54. Laxenburg, Austria: IIASA.

Eyre, S. R. (1968). *Vegetation and soils. A World Picture*. London: Edward Arnold.

Forest Resources. The State of Environment 1985, pp. 119-33. (1985), Paris: ECDE.

Hoffman, G. (1985). Die potentielle natürliche Nettoprimärproduktion an oberirdischer Pflanzen-trockenmasse - ihre Quantifizierung und Kartierung für das Gebiet der DDR. *Beiträge für die Forstwirtschaft 19*, 110-16.

Kandler, O. (1992). Historical declines and diebacks of Central European forests and present conditions. *Toxicology and Chemistry 11*, 1077-92.

Kauppi, P. E., Mielikäinen, K. & Kuusela, K. (1992). Biomass and carbon budget of European forests, 1971 to 1990. *Science 3*, Vol. 256, April 1992, pp. 70-4.

Kuusela, K. (1968). Growing stock management in Central European and in British forestry. *Communicationes Instituti Forestalis Fenniae 66.2*. Helsinki.

Kuusela, K. (1984). European forest resources and their use. *Finnish National Fund for Research and Development, Series B*, No. 75. Helsinki (in Finnish).

Kuusela, K. (1985). European forest resources and the trade of industrial wood in 1950-2000, *Finnish National Fund for Research and Development, Series B*, No. 79. Helsinki (in Finnish).

Kuusela, K. (1987). Silvicultural regimes in the cause and effect relationships of the forest damage situation in Central Europe. WP-37-31. Laxenburg, Austria: IIASA.

Kuusela, K. (1990). The dynamics of boreal coniferous forests. *Finnish National Fund for Research Development*, 112. Helsinki.

Mayer, H. (1984). *Wälder Europas*. Stuttgart: Gustav Fischer Verlag.

Paterson, S.S (1956). The forest area of the world and its potential productivity. *Meddelande från Götegorgs Universitets Geografiska Institution* 51. Göteborg, Sweden.

UN (1964). *European Timber Trends and Prospects. A New Appraisal 1950-1975.* New York: UN-ECE/FAO.

UN (1976). *European Timber Trends and Prospects 1950 to 2000.* Geneva: UN-ECE/FAO.

UN (1985). *The Forest Resources of the ECE Region (Europe, the USSR, North America).* Geneva: UN-ECE/FAO.

UN (1986). *European Timber Trends and Prospects to the Year 2000 and Beyond,* Volumes I and II. New York: UN-ECE/FAO.

UN (1987). Forestry and the Forest Industries: Past and Future, ed. E. G. Richards. *Forestry Sciences* No. 27. Dordrecht, The Netherlands: ECE/FAO.

UN (1992). *The Forest Resources of the Temperate Zones,* Volumes 1 and 2. New York: UN-ECE/FAO.

Windhorst, H.-W. (1978). *Geographie der Wald-und Forstwirtschaft.* Stuttgart: B. G. Teubner.